AF718366

WITHDRAWN

Monographs in Electrical and Electronic Engineering

Series editors: P. Hammond, T. J. E. Miller, and S. Yamamura

Monographs in Electrical and Electronic Engineering

10. *The theory of linear induction machinery* (1980) Michel Poloujadoff
12. *Energy methods in electromagnetism* (1981) P. Hammond
15. *Superconducting rotating electrical machines* (1983) J. R. Bumby
16. *Stepping motors and their microprocessor controls* (1984) T. Kenjo
17. *Machinery noise measurement* (1985) S. J. Yang and A. J. Ellison
18. *Permanent-magnet and brushless d.c. motors* (1985) T. Kenjo and S. Nagamori
19. *Metal–semiconductor contacts: Second edition* (1988) E. H. Rhoderick and R. H. Williams
20. *Introduction to power electronics* (1988) Eiichi Ohno
21. *Brushless permanent-magnet and reluctance motor drives* (1989) T. J. E. Miller
22. *Vector control of a.c. machines* (1990) Peter Vas
23. *Brushless servomotors: fundamentals and applications* (1990) Y. Dote and S. Kinoshita
24. *Semiconductor devices, circuits, and systems* (1991) Albrecht Möschwitzer
25. *Electrical machines and drives: a space-vector theory approach* (1992) Peter Vas
26. *Spiral vector theory of a.c. circuits and machines* (1992) Sakae Yamamura
27. *Parameter estimation, condition monitoring, and diagnosis of electrical machines* (1993) Peter Vas
28. *An introduction to ultrasonic motors* (1993) S. Sashida and T. Kenjo
29. *Ultrasonic motors: theory and applications* (1993) S. Ueha and Y. Tomikawa
30. *Linear induction drives* (1993) J. F. Gieras
31. *Switched reluctance motors and their control* (1993) T. J. E. Miller
32. *Numerical modelling of eddy currents* (1993) Andrzej Krawczyk and John A. Tegopoulos

Numerical Modelling of Eddy Currents

Andrzej Krawczyk

Institute of Electrical Engineering
Warsaw

and

John A. Tegopoulos

Department of Electrical Engineering
National Technical University of Athens

CLARENDON PRESS · OXFORD
1993

TK
2271
.K73
1993

Oxford University Press, Walton Street, Oxford OX2 6DP
Oxford New York Toronto
Delhi Bombay Calcutta Madras Karachi
Kuala Lumpur Singapore Hong Kong Tokyo
Nairobi Dar es Salaam Cape Town
Melbourne Auckland Madrid
and associated companies in
Berlin Ibadan

Oxford is a trade mark of Oxford University Press

Published in the United States by
Oxford University Press Inc., New York

© Andrzej Krawczyk and John A. Tegopoulos, 1993

All rights reserved. No part of this publication may be reproduced, stored in a retrieval system, or transmitted, in any form or by any means, without the prior permission in writing of Oxford University Press. Within the UK, exceptions are allowed in respect of any fair dealing for the purpose of research or private study, or criticism or review, as permitted under the Copyright, Designs and Patents Act, 1988, or in the case of reprographic reproduction in accordance with the terms of licences issued by the Copyright Licensing Agency. Enquiries concerning reproduction outside these terms and in other countries should be sent to the Rights Department, Oxford University Press, at the address above.

A catalogue record for this book is available from the British Library

Library of Congress Cataloging in Publication Data
(Data available on request)
ISBN 0–19–859382–1

Typeset by the authors using LaTeX
Printed in Great Britain on acid-free paper by
Biddles Ltd, Guildford and King's Lynn

PREFACE

If you cannot express something numerically
your knowledge is poor and insufficient.

Sir William Thomson

During the last three decades great progress has been made in developing and using numerical methods for solving problems in electromagnetic fields at low and high frequencies. Undoubtedly this progress was made possible by the advent of the digital computer. Among these problems, many involve eddy currents. Such problems were previously tackled using analytical solutions, but these are limited to simple geometrical configurations such as planes, cylinders, or spheres. In addition, the problems have to be linear. Such configurations can be found in technical applications, but the majority of electromagnetic devices today do not have a simple geometry. These cases can be handled exclusively by using numerical solutions and this is the reason for their extensive use.

An extensive literature has been produced so far consisting of papers published in technical journals or presented at conferences. Also, several computer program packages are commercially available.

The authors feel that this knowledge should be unified in a systematic way so that students and researchers may have a concise reference to resort to, and this was the motivation for writing the book.

Although they have the disadvantage of being limited to simple geometries, analytical solutions have the merit that they are fast and accurate, and maintain contact with the physical picture. However, numerical solutions can handle practically any configuration, regardless of how complicated its boundaries may be. It is this feature that renders numerical solutions indispensable. Of course, the fact that these solutions are iterative results in long computing times, and contact with the physical picture is practically lost.

The effects of eddy currents are assuming more and more importance, following the trend of designing electromagnetic devices with higher power density. This requires from engineers a considerably greater knowledge of electromagnetic phenomena when designing and operating such devices, since eddy currents directly influence the performance of electrical apparatus. This influence may be desirable or undesirable. Thus, there exist devices which are based on the development of eddy currents, such as linear motors or launchers, devices for magnetic forming, for the generation or concentration of high magnetic fields, and so on. In these cases, eddy currents are desirable and exploited. In other cases eddy currents are undesirable as they cause Joule losses and should be minimized. In both cases they need to be accurately described and determined. The determination of eddy currents in a given configuration by using numerical solutions is the scope of this book. Specifically, the methods available with appropriate criteria for their use are presented. Procedures of solutions are also given, along with some typical applications.

In its comprehensive treatment of the numerical solution of eddy current problems, this book is oriented towards giving mathematical simulations for existing methods, but it also discusses the way that pertinent equations should be discretized. However, it does not cover the discretization itself. Relatively lengthy discussions are provided about the physical connotations of the methods or problems.

We therefore feel that the book is especially appropriate for young engineers and researchers, and could serve as a manual for the methodology and related physical background.

A rich bibliography is included which will be of considerable help for further reading in carrying out solutions of real problems and the study of their results.

Warsaw Andrzej Krawczyk
Athens John A. Tegopoulos
August 1993

CONTENTS

1

INTRODUCTION

Eddy currents have long been of interest to electrical engineers, and have been distinguished as a separate entity in the field of electromagnetism since Foucault[1] first discovered their nature. Indeed, no one who deals with electrical engineering, either theoretically or practically, can be dispassionate about eddy currents. This results from their dichotomical nature, that is, they can be either desired or undesired. In one field of applied electricity they are consciously generated, and in others they are injurious and should be avoided or reduced. In fact, eddy currents are one of the main problems encountered in designing electrical apparatus, and for this reason we have focused our attention on this narrow topic.

Leaving aside all the general electromagnetics books which tend not to describe eddy currents in detail, or those books which are not accessible to a wide readership because they are not published in English, there are three books dealing with eddy currents *per se* up to now (Lameraner and Stafl 1968; Stoll 1974; Tegopoulos and Kriezis 1985). One should mention here the book by Bunet (1933) but this describes only the phenomenology of eddy currents, so that it is useless from a computational point of view. The book by Tegopoulos and Kriezis is concerned with analytical methods applied mainly to two-dimensional configurations. The book by Stoll 1974 is concerned with both analytical and numerical methods, giving a wide spectrum of eddy current considerations. From the point of view of the numerical approach the finite difference method has been widely documented there. The book by Lammeraner and Stafl is devoted to the analytical approach only and covers one-dimensional configurations, so that its usefulness is limited.

The scope of this book is such that it could be considered as a prescription for the numerical modelling of eddy current phenomena. At this point some comments should be given explaining what such a modelling really means. The following definition of such a model is suggested here:

A mathematical model is a set of mathematical relations which describes a physical phenomenon governed by the fundamental and constitutive physical laws, the description being made in a unique, coherent and stable way.

[1] Jean Bernard Leon Foucault (1819–1868), a French physicist, discovered eddy currents (Foucault's currents) induced in a copper disc rotating in high magnetic field.

The term *unique* means that for one set of adequate data the model gives one and only one solution.

The model is *coherent* if the variables employed for every model are compatible.

The model is *stable* if it is not sensitive to small perturbations, that is, perturbations of data and/or parameters of the model cause proportional changes in the solution.

To build the model, defined as above, three questions shold be considered:

- What is the most appropriate field quantity describing a given electromagnetic phenomenon?
- What is the mathematical formulation of this phenomenon using the above quantity?
- What is the approximation connected with this formulation?

In Chapter 2 the physical and mathematical background is developed on which the different modelling methods are based.

The following chapters (3 and 4) attempt to show the possible answers to above questions for different applications. Both continuous and numerical models are described there and numerical methods are given, followed by examples of applications.

Chapter 5 describes the use of numerical methods for special cases encountered in engineering practice, which are important from a theoretical point of view. They are known as the 'open boundary problems' and the interface problems.

Basically, we shall be concerned with eddy currents induced in magnetically linear conducting media. However, eddy currents may also appear in non-linear media (ferromagnetic). This case is covered in Chapter 6 where corresponding methods are described.

Finally, Chapter 7 considers eddy currents related to *moving media*. To this last case belongs the application of rotating and linear motors.

The authors are aware that many problems have not been included in this book since, although the subject is narrow, the current stream of scientific activity cannot be covered by only one publication. This book aims at the methodology rather than at covering the entire state of the art.

At the end of the introductory remarks let us notice that the whole activity in general field problems is confined to the paradigm (in Kuhn's (1962) sense) which is, briefly speaking, the methodology following Maxwell's, Ampere's, Faraday's, and other laws. This methodology keeps all scientists confined to the set of notions and approaches determined by these laws and the book also complies with them. It is quite clear that it is not going to be changed in the near future. However, some attempts have been made in this direction and one of them is the field analysis using differential geometry (that is, geometry of field) instead of the differential analysis and geometry of object (Hammond 1989). The results of such attempts cannot be applicable, unless the capacity of existing computers reaches a limit.

2

METHODS OF EDDY CURRENT ANALYSIS

2.1 Governing equations

All electromagnetic phenomena are governed by the well-known laws, rules and formulae which are described in existing relevant books, so that it would seem to be unnecessary to repeat them here. Nevertheless, it would be difficult to publish a book on eddy currents without giving a brief summary of these laws. As it was mentioned in the Introduction it was Foucault who distinguished eddy currents from the general notion of electric current. However, eddy currents are defined in the general theory of electromagnetism. In fact, they can be deduced from Faraday's law as well as from Maxwell's equations, although Maxwell himself did not focus his attention on this problem. Nevertheless, the eddy current problem can be considered as a result of electromagnetic induction but with the electromotive force induced in a massive conductor instead of a circuit made of wires or filaments.

Before quoting the field equations, the following assumptions are made, normally used in eddy current analysis:

- the displacement current is neglected;
- conducting media are isotropic;
- conducting objects are treated macroscopically (no atomic effects are considered);
- the structure of media is homogeneuous (no domain theory is considered);
- the thermal non-linearity is not considered (the conductivity is independent of field values);
- the magnetic permeability μ of the conducting media is constant; this assumption is valid for all Chapters except Chapter 6 which briefly describes non-linear problems.

It seems that some comment on the first assumption is required. There is no doubt that this assumption is correct when dealing with low-frequency harmonic fields. The question arises when the variation in time is substantially high or when one deals with transient phenomena. But even then the assumption that the displacement current vanishes is valid as long as the dimension of the object analysed is small and its conductivity is high enough (as with copper or aluminium). It was mathematically proved by

Grumet (1959) that for a field excitation in the form of a unit step function the displacement current in the conducting half-space can be ignored. An exhaustive analysis of this problem has been presented in the book mentioned in the Introduction (Tegopoulos and Kriezis 1985) as well as in others (for example, Schieber 1986) where the criteria for the validity of the assumption are presented.

Under the above assumptions the equations describing eddy currents represent fundamental laws of electromagnetism and are as follows:

$$\operatorname{curl} \boldsymbol{H} = \boldsymbol{j} \qquad \text{Ampere's law} \tag{2.1}$$

$$\operatorname{curl} \boldsymbol{E} = -\frac{\partial \boldsymbol{B}}{\partial t} \qquad \text{Faraday's law} \tag{2.2}$$

$$\operatorname{div} \boldsymbol{B} = 0 \qquad \text{Gauss's law} \tag{2.3}$$

$$\boldsymbol{B} = \mu \boldsymbol{H} \qquad \text{constitutive relation} \tag{2.4}$$

$$\boldsymbol{j} = \gamma \boldsymbol{E} \qquad \text{constitutive relation (Ohm's law)} \tag{2.5}$$

where $\boldsymbol{H}$, $\boldsymbol{E}$ are the magnetic and electric field strength, respectively, $\boldsymbol{B}$ is the magnetic flux density and $\boldsymbol{j}$ is the current density, μ and γ are the magnetic permeability and electric conductivity, respectively.

The above set of equations is complemented by the interface conditions which ensure that the magnetic field solution is unique in the case of non-homogeneity of the region in question. They result from the physical features of the electromagnetic field and are as follows:

$$\boldsymbol{n} \cdot (\boldsymbol{B}_1 - \boldsymbol{B}_2) = 0 \tag{2.6}$$

$$\boldsymbol{n} \times (\boldsymbol{H}_1 - \boldsymbol{H}_2) = 0 \tag{2.7}$$

where '1' and '2' denote two sub-regions on the interface of which the above conditions hold,[1] and $\boldsymbol{n}$ is the unit vector normal to the interface. The interface conditions are called boundary conditions when the region considered is surrounded by an environment which enforces the field behaviour on its boundary. This may be, for example, highly permeable material in contact with a non-permeable environment. In this case the tangential component of the magnetic field strength tends to zero. In another case a highly conducting material may be in contact with non-conducting environment. Then the normal component of magnetic flux density vanishes.

Eliminating $\boldsymbol{E}$ and $\boldsymbol{j}$ from the fundamental equations, one can obtain one equation which, when considering all parameter variations, is as follows:

$$\nabla^2 \boldsymbol{H} = \gamma \frac{\partial B}{\partial H} \frac{\partial \boldsymbol{H}}{\partial t} - \operatorname{grad}(\boldsymbol{H} \cdot \frac{\operatorname{grad} \mu}{\mu}) - \frac{\operatorname{grad} \gamma}{\gamma} \times \operatorname{curl} \boldsymbol{H} \tag{2.8}$$

[1] The second condition is under the assumption that there is no surface current and it is always fulfilled in low frequency and finite conductivity problems

Of course, in real cases under the above assumptions eqn (2.8) is considerably simplified since all the material parameters are constant both in time and in space. Thus, we have

$$\nabla^2 \boldsymbol{H} = \gamma\mu \frac{\partial \boldsymbol{H}}{\partial t} \tag{2.9}$$

For sinusoidal time variation one can make the Fourier's transformation $\boldsymbol{H}(t) = Re(\hat{\boldsymbol{H}} \exp(i\omega t)$, where ω is the angular frequency, $i = \sqrt{-1}$ and in terms of phasors, we have

$$\nabla^2 \hat{\boldsymbol{H}} = i\omega\mu\gamma \hat{\boldsymbol{H}} \tag{2.10}$$

Let us make here some terminological digression. Throughout the book the following time states will be distinguished:

- *transient* for an arbitrary time dependence;
- *quasi-stationary* for the sinusoidal time dependence;
- *static* for constant in time problems.

The above equations are valid for the bodies at rest. Needless to add that the majority of eddy current applications involve moving media. Indeed, eddy currents can be also created by motion and then the basic equation includes an additional term as follows:

$$\nabla^2 \boldsymbol{H} + \mu\gamma \operatorname{curl}(\boldsymbol{v} \times \boldsymbol{H}) = \mu\gamma \frac{\partial \boldsymbol{H}}{\partial t} \tag{2.11}$$

where v is the relative speed between the body and the source field.

Assuming $\boldsymbol{v}$ =const in eqn (2.11) the formula can be simplified to:

$$\nabla^2 \boldsymbol{H} - \mu\gamma(\boldsymbol{v} \cdot \operatorname{grad})\boldsymbol{H} = \mu\gamma \frac{\partial \boldsymbol{H}}{\partial t} \tag{2.12}$$

Although the nature of eddy currents is unique, eddy currents can be produced by two mechanisms of induction as follows:

- eddy currents induced by time-varying excitation and represented by the term $\partial \boldsymbol{H}/\partial t$;
- eddy currents induced by motion and represented by the term $\operatorname{curl}(\boldsymbol{v} \times \boldsymbol{H})$.

As in fluid mechanics one can introduce here the *magnetic Reynolds number*

$$R_m = \gamma\mu L v \tag{2.13}$$

where L is the characteristic length of the moving body (Moon 1984).

This enables us to evaluate which type of eddy currents dominates. If R_m is small the first type prevails while for R_m large the effect of motion dominates. For R_m being of the order unity neither the first nor the second effect can be ignored. This case mostly occurs when analysing the induction pumping of liquid metals. Numerical aspects of the problems involving motion are discussed in Chapter 7 and until then all material developed concerns eddy currents induced by time-varying fields.

Eddy currents result directly from Faraday's law (see eqn (2.2)) and generally have three components in space. However, in many technical cases the problem of eddy currents can be considered in two-dimensional space. In such cases two situations are discussed:

- *H-excitation,*
- *j-excitation.*

In the first case the magnetic field strength has one component only while the eddy current density is expressed by two components. In the second case the eddy current density has one component and the magnetic field strength has two. The latter, however, is usually described in terms of the magnetic vector potential which has also one component.

In practice, one looks not for eddy currents themselves but for their effects and two of them are of special importance. These are: power (ohmic) losses and electrodynamic forces. The latter quantity is necessary for analysing electromechanical converters and is described by the formula

$$\boldsymbol{f} = \boldsymbol{B} \times \boldsymbol{j} \tag{2.14}$$

while the first one is required when designing electrical machines and transformers.

At this point we introduce Poynting's theorem[2] and Poynting's vector which expresses the energetic balance in the electromagnetic field and is given by

$$\boldsymbol{S} = \boldsymbol{E} \times \boldsymbol{H} \tag{2.15}$$

Integrating this vector and using the basic equations re-arranged with vector identities one obtains

$$\oint_{\Gamma} (\boldsymbol{E} \times \boldsymbol{H}) \cdot \boldsymbol{n} d\Gamma = -\int_{\Omega} \boldsymbol{H} \frac{\partial \boldsymbol{B}}{\partial t} d\Omega - \int_{\Omega} \boldsymbol{E} \cdot \boldsymbol{j} d\Omega \tag{2.16}$$

where Ω is the eddy current space, Γ its boundary and $\boldsymbol{n}$ the unit vector of normal outward direction on Γ.

The second integral of the right-hand side represents just power losses (ohmic or Joule's) while the first one represents magnetic energy stored in

[2]Poynting, J. H. (1884). On the transfer of energy in the electromagnetic field. *Transactions of Royal Society* Part II, 343-58.

volume. Vector $\boldsymbol{S}$ itself pictures the energy density influx to the eddy current body.[3] Poynting's vector is especially useful for harmonic (sinusoidal) problems. Then the Poynting's vector assumes the form:

$$\hat{\boldsymbol{S}} = \frac{1}{2}\hat{\boldsymbol{E}} \times \hat{\boldsymbol{H}}^* \tag{2.17}$$

where $\hat{\boldsymbol{H}}^*$ is the conjugate value to $\hat{\boldsymbol{H}}$. The Poynting theorem in phasor terms has the form

$$\oint_\Gamma \hat{\boldsymbol{S}} \cdot \boldsymbol{n} d\Gamma = -i2\omega \int_\Omega \frac{1}{4}\mu_0\mu_r \mid \hat{\boldsymbol{H}} \mid^2 d\Omega - \frac{1}{2\gamma}\int_\Omega \mid \hat{\boldsymbol{j}} \mid^2 d\Omega \tag{2.18}$$

Using the above formula one can easily see that the power losses (AC resistance) are the real part of the influx of vector $\hat{\boldsymbol{S}}$ while the magnetic energy, hence AC inductance, is given by the imaginary part of it. Since the integral is of boundary nature it has the practical consequence that boundary values determine energy for the whole system.

2.2 Survey of methods

Since eddy currents have always been drawing the attention of researchers and engineers, their determination was of interest from the very beginning of electrical engineering. The problem was recognized very often as a skin-effect problem and referred to in the literature by this name. Who should be mentioned as a founder of eddy current calculation? It is difficult to state it precisely but it seems that the first valuable results were obtained by A.B. Field,[4] who calculated the slot leakage impedance of electrical machines by means of the skin-effect analysis. The importance of his work is confirmed by the fact that Field's formulae are still being taught and apllied in industry. Of course one can give an account of the work of many other scientists who attempted to tackle the problem, but they are shadowed by Field's work.

Next, one can quote here the work of Schwenkhagen (1927) who solved the problem of eddy currents in a conductor with rectangular cross-section. Indeed, it was probably the first two-dimensional case solved. He used the method of integral equation although not entirely consciously. Namely, his method consisted in dividing a conductor into small axial elements and then establishing a set of mutual inductances. This led in fact to a system of linear equations, which is the same as the one resulting from the integral equation treated by the collocation method. The usefulness of the integral equation method in describing eddy current phenomena has been confirmed

[3] Note that vector S has a sense if eddy currents exist. In magnetostatic problems the vector product of E and H is the product only and has no physical meaning

[4] Field, A. B. (1905). Eddy current in large slot wound conductors. *AIEE*, 761-788.

by the paper by Gross (1940). He used in turn the method of successive iterations, which obviously results from solving an integral equation by means of the successive approximations method (Neumann's series).

In the meantime some researchers were using Fourier's (separation of variables) method applied to the differential equation. It is interesting to see that, since the differential equation approach required boundary conditions to be prescribed, the corresponding solutions were applicable to this particular problem only and so lacked generality. This did not apply to integral solutions. An excellent survey of the analytical solutions obtained from Fourier's separation of variables or transformation methods is given in the books already mentioned. Since about 1950 the numerical methods began to play an important role in this subject: first, the finite difference method, then the finite element method, and lately since about 1980 the boundary element method. Actually these approaches constitute the subject of this book and they will be dealt with later.

The short review given above was devoted to quasi-stationary problems that is, problems with sinusoidal variation of time. Slightly different was the development of solutions in transient problems. Such solutions were more difficult than static ones and therefore they appeared considerably later. For the analysis in time no simple methods could be used. The essential turning point in this field was made by Heller (1954), Grumet (1959), Chura (1976) and others who applied the Laplace transformation method in order to solve the time-dependent problem. Indeed, such an approach appeared to be very profitable and fruitful and was successfully used until the 1970s (Krawczyk 1980). Also the idea of the Laplace transformation was used by some researchers who took the solution of harmonic eddy currents and, substituting Fourier's operator '$i\omega$' by the Laplace one 's', searched for the inverse transformation of the operator solution. In other words, they looked for the time constant of the transient process which immediately resulted from the operator solution (Mocanu 1971). Of course, some of the transient problems were solved by the method of separation of variables and again an extensive survey of these solutions is given in the books mentioned above. It should be stressed that all analytical solutions of transient problems, as well as the quasi-stationary ones, were obtained for relatively simple geometric configurations including the one-dimensional ones.

Besides the assumptions, commonly used when analytical solutions are considered, the asymptotic states of eddy currents were introduced. Following Hammond (1971) and Stoll (1974) one can distinguish

- resistance-limited state,
- inductance-limited state.

The first one describes the situation when eddy currents do not affect the magnetic field and thus are limited by the resistance of the body. The

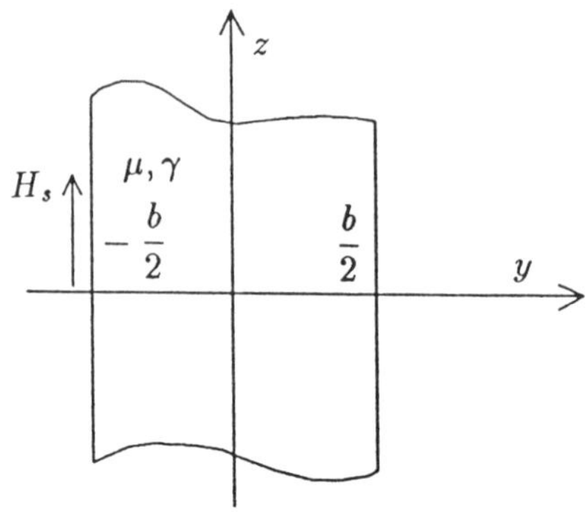

FIG. 2.1. Conducting plate

second case is applied when eddy currents develop in the body until they vanish. Then the inductance limits them. It is obvious that the use of both states is determined by geometry of the body and parameters of the exciting field as well.

Let us now consider an example. The one-dimensional bounded conducting plate is placed in the sinusoidal magnetic field (Fig. 2.1) For the model shown in Fig. 2.1 both eddy currents and magnetic field strength have one component only, that of x and z direction, respectively. Then solving eqn (2.9) leads to the following formula for the eddy current density:

$$\hat{j} = \alpha \hat{H}_s \frac{sinh\alpha y}{cosh\alpha(b/2)} \tag{2.19}$$

where $\alpha = \sqrt{i\omega\mu\gamma} = (1+i)/\delta$, δ is the so-called skin depth.

In the resistance-limited case eddy currents do not influence the magnetic field inside the plate and thus the power losses can be calculated from

$$P_e = \frac{|\hat{H}_s|^2}{\gamma\delta}\frac{4}{3}\frac{b^3}{\delta^3} \tag{2.20}$$

This formula may be used if $b/\delta \ll 1$.

In the inductance-limited case one can assume that the magnetic field inside the plate is determined by eddy currents and then we have

$$P_e = \frac{|\hat{H}_s|^2}{\gamma\delta} \tag{2.21}$$

This formula applies when $b/\delta \gg 1$.

If neither of the two conditions is fulfilled the exact formula has to be used which is as follows:

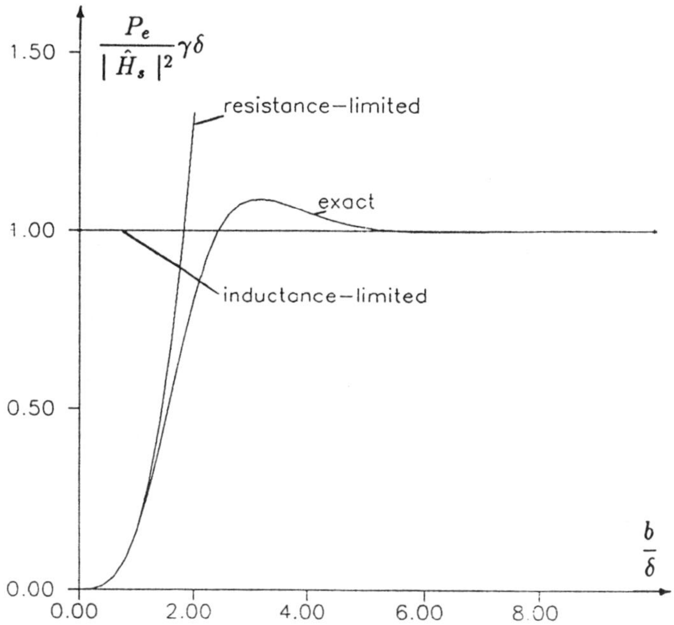

FIG. 2.2. Relative power losses as function of b/δ

$$P_e = \frac{|\hat{H}_s|^2}{\gamma\delta}\frac{\sinh(b/\delta) - \sin(b/\delta)}{\cosh(b/\delta) + \cos(b/\delta)} \tag{2.22}$$

The graph of this functions and those of asymptotic states is given in Fig. 2.2.

All above considerations concern the problem of the so called 'stiff sources', that is, the cases when the magnetic source is not affected by eddy current field. If this were not so, the problem could not be treated as simply as eqn (2.22) suggests.

One can quote here the two practical cases which are representative of the above asymptotic states. It is the sheet of the transformer core which is always treated as the resistance-limited case. On the other hand the wall of the transformer tank or other massive conductive elements of electric machines can be considered as a plate of infinite thickness and then inductance-limited approximation is adopted. That is why the two asymptotic states were intuitively used by engineers long ago (for example, all the classical formulae for eddy current power losses in a transformer core) and the classification given above which follows that of Stoll and Hammond just confirms well-known technical approaches.

The numerical treatment of resistance-limited cases does not involve any serious difficulty. On the contrary situations approaching the inductance-limited case ($b/\delta > 1$) may lead to serious discretization problems as the mesh dimension near the conductor surface must be appreciably smaller than the skin depth in order not to lose the information on near-surface eddy currents. Such a constraint implies a great number of unknows or may lead to numerical problems which are ill-conditioned (see also Section 4.3.2, p. 59).

The extreme inductance-limited situation, that is, $b/\delta \gg 1$, leads to the notion of the impedance boundary condition (IBC) (Fawzi *et al.* 1985). Using such an asymptotic state, one can relate the surface vectors $\boldsymbol{H}$ and $\boldsymbol{E}$ by the surface impedance $\mathcal{Z}$:

$$\boldsymbol{n} \times \boldsymbol{E} = \mathcal{Z}\boldsymbol{n} \times (\boldsymbol{n} \times \boldsymbol{H}) \tag{2.23}$$

where

$$\mathcal{Z} = \frac{(1-i)\mu\omega\delta}{2} \tag{2.24}$$

Introducing the above conditions, one has to solve the field problem in the non-conducting region only, that is, solve Laplace's or Poisson's equation, which is much simpler than solving eddy current problem. The IBC simplifies itself in two-dimensional or one-dimensional problems and has various forms depending on the variables used. It can be proved that for high-frequency problem ($\delta \to 0$) eqn (2.23) shows that the normal component of $\boldsymbol{H}$ vanishes, which is the well-known feature of perfect conducting screens. The impedance boundary condition is especially convenient for the boundary-integral formulation of static fields.

Recapitulating, one can say that in the past various methods were used in solving eddy current problems and all of them were more or less satisfactory. However, modern technology requires the calculation of eddy currents in more complex configurations, a fact which motivates analysts to look for more appropriate solutions. Nevertheless, analytical solutions are still useful for another reason: they can serve as model (test) solutions to validate numerical results.

3

CONTINUOUS MODEL OF EDDY CURRENTS

3.1 The quantities describing eddy current problems

The eddy current phenomenon can be expressed in terms of certain variables which, in turn, correspond to the various ways of analysis. In electromagnetism the variables describing the field are of two kinds. The first type of variables concerns the field itself while the second type treats the field indirectly by using mathematical entities, known as potentials. Both type have advantages and disadvantages and they should be used for each case as appropriate. The main benefit of using the field variables results from the fact that they allow observing the field as a physical reality. The disadvantage is that solving the problem requires considerably more work. On the other hand, introducing potentials gives an analyst the possibility of less work for the solution of the problem in question. In contrast, potentials are of mathematical rather than physical nature, so that this analysis is far from reality and it is not possible to use here physical intuition. In analysing eddy currents, because of their nature, the main potentials used are vectorial although scalar potentials are also considered as complementary to the vectorial ones.

3.1.1 *Field variables*

The main field quantities employed are the vectors $\boldsymbol{H}$, $\boldsymbol{B}$, and $\boldsymbol{E}$. They are magnetic field strength, magnetic flux density, and electric field strength, respectively. These variables follow directly from the fundamental electromagnetic laws and carry an important physical meaning, which is an advantage. However, in practice they can be used only in the fundamental or constitutive equations, which from the calculation point of view, is not particularly beneficial. For example, to solve an eddy current problem we have to use at least three equations for the three-dimensional case and two equations for the two-dimensional case. There are of course some simple and symmetrical cases in which one can reduce the number of equations but this occurs rarely. Nevertheless we can use these variables to better understand the field nature. What is more, the mathematical analysis gives field variables directly, without additional differentiation as is the case with potentials. Also the modern three-dimensional analysis of eddy currents reveals the benefits of using the direct field quantities. This will be discussed later on at length (see Chapter 4). Yet, on the basis of vectors $\boldsymbol{B}$ and

$\boldsymbol{H}$, one can attempt to describe in a mathematical way the dependence between $\boldsymbol{B}$ and $\boldsymbol{H}$ for non-linear cases, which seems to be very natural. The physics of the magnetic non-linearity is just given in terms of the B–H curve. This is well-described in a great number of books (for example, Bozorth 1965), and therefore we shall not discuss it here. One comment should be added: that even for the linear case the vectors $\boldsymbol{B}$ and $\boldsymbol{H}$ are of different nature. Thus vector $\boldsymbol{H}$ assigned to every point of the field acts as the source of the field. On the other hand vector $\boldsymbol{B}$ contains information on the physical and geometric properties of the material. From this point of view vector $\boldsymbol{B}$ is more significant physically.

The classical problem requiring the field variables is the one-dimensional analysis of field penetration into a massive slab with given constant values of permeability μ and conductivity γ. For example, taking the conducting medium of infinite thickness in the direction of field penetration (Fig. 3.1) and assuming that the field H_s on the metal - air interface is uniform (plane-parallel wave) one gets the following solutions for eddy current distributions in two cases:

(1) quasi-stationary with an angular frequency ω:

$$\hat{j}(x) = \sqrt{\frac{\omega\mu}{2\gamma}}\hat{H}_s \exp(-\alpha x) \tag{3.1}$$

where $\alpha = (1+i)\sqrt{\omega\mu\gamma/2}, i = \sqrt{-1}$

(2) transient:

$$j(x,t) = \int_0^t H_s(\tau)(t-\tau)^{-5/2} \exp(-\alpha^2 x^2/4(t-\tau))d\tau \tag{3.2}$$

where $\alpha^2 = \mu\gamma$ and in the case $H_s(t) = H_0 1(t)$ (unit-step function) the formula is much simpler:

$$j(x,t) = \sqrt{\frac{\mu\gamma}{\pi t}} H_0 \exp\left(-\frac{\alpha^2 x^2}{4t}\right) \tag{3.3}$$

Although the example quoted above seems to be simple, many researchers during the last few decades (Heller 1953; Chura 1976; Krawczyk 1980) examined it carefully, especially in the transient case, mainly because it gives information about the behaviour of power losses or other integral results, such as forces and inductances. An interesting complement of the example would be to introduce a non-linearity and this was done by Haas (1978).

Moreover, this example introduces the reader to the mathematical expressions of eddy currents as well as showing the commonly used notions

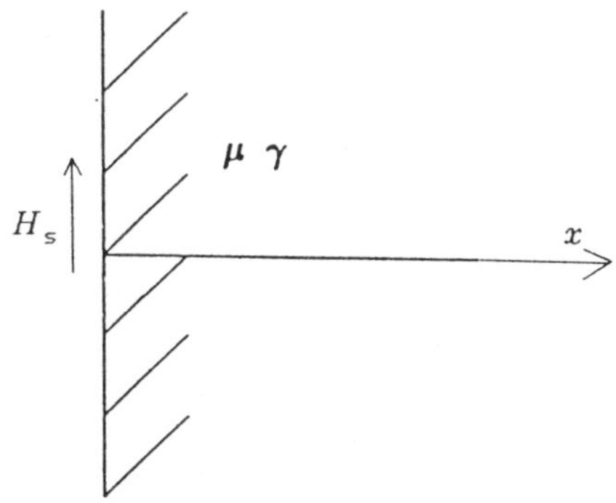

FIG. 3.1. Half-space excited by homogeneous magnetic field

of this analysis such as the skin depth (already mentioned in the previous chapter) which has the form

- for the quasi-stationary case:

$$\delta = \sqrt{\frac{2}{\omega\mu\gamma}} \tag{3.4}$$

- for the transient case:

$$\delta(t) = \sqrt{\frac{4t}{\gamma\mu}} \tag{3.5}$$

This very simple, even trivial example has, however, some technically important connotations as it allows us to link material and field parameters and to make an initial estimate of eddy current effects (power losses, temperature, and so on).

The direct variables are also useful in the two-dimensional case, when considering special circumstances of the problem in question. Namely, if one assumes that the one component of $\boldsymbol{H}$ or $\boldsymbol{B}$ exists and two components of eddy currents are expected, the field equation is as follows:

$$\frac{\partial^2 H}{\partial x^2} + \frac{\partial^2 H}{\partial y^2} = \mu\gamma\frac{\partial H}{\partial t} \tag{3.6}$$

and then eddy current components are of the form

$$j_x = \frac{\partial H}{\partial y} \tag{3.7}$$

$$j_y = -\frac{\partial H}{\partial x} \tag{3.8}$$

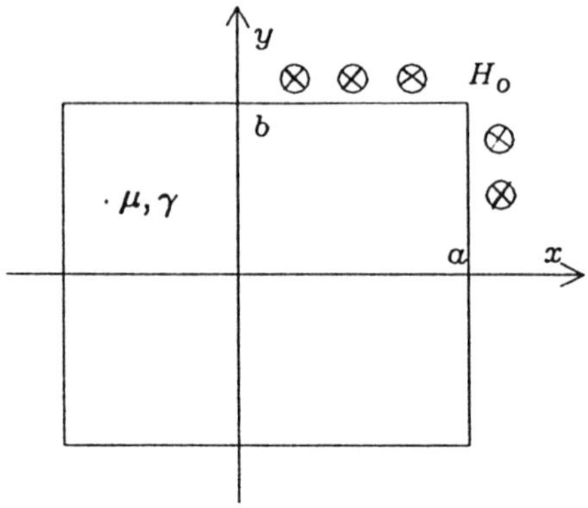

FIG. 3.2. Two-dimensional arrangement

Physically, this case implies the previously mentioned *flux excitation*, that is, the long conducting medium is subjected to a field parallel to its long direction (Fig. 3.2).

If the cross-section of the conductor is geometrically simple (for example, rectangular) the case can also have an analytical solution both for quasi-stationary and transient problems. Let us say that the cross-section is the rectangle with the sides (-a,a) and (-b,b). Then the analytical solutions of H (linear case) are as follows:

$$\hat{H}(x,y) = \hat{H}_0 \left(\frac{\cosh \alpha x \cosh \alpha y}{\cosh \alpha a \cosh \alpha b} \right) \tag{3.9}$$

where $\alpha = (1+i)\sqrt{\mu\omega\gamma/2}$.

$$H(x,y,t) = H_0 - H_0 \frac{16}{\pi^2} \sum_{n=1}^{\infty} \sum_{m=1}^{\infty} C_{n,m} \cos(P_n x) \cos(Q_m y) e^{-\lambda_{n,m}^2 t} \tag{3.10}$$

where

$$\begin{aligned} C_{n,m} &= \frac{(-1)^{n+m}}{(2n-1)(2m-1)} \\ P_n &= \frac{2n-1}{2a}\pi \\ Q_m &= \frac{2m-1}{2b}\pi \\ \lambda_{n,m}^2 &= \frac{P_n^2 + Q_m^2}{\mu\gamma} \end{aligned}$$

For the three-dimensional case, in turn, the use of direct variables is not so simple, since the scalar potential cannot be omitted then. Following consequently fundamental equations (2.1–2.3) and using the constitutive relations (2.4–2.5) one obtains:

$$\nabla^2 \boldsymbol{H}^+ = \mu\gamma \frac{\partial \boldsymbol{H}^+}{\partial t} \tag{3.11}$$

$$\nabla^2 \Phi^- = 0 \tag{3.12}$$

where $\boldsymbol{H}^- = -\operatorname{grad}\Phi^-$.

The two regions: exterior one denoted '−' and interior '+' are coupled by interface conditions, which are as follows:

$$\boldsymbol{n}\cdot(\mu\boldsymbol{H}^+ - \mu\operatorname{grad}\Phi^-) = \mu\boldsymbol{n}\cdot\boldsymbol{H}_e \tag{3.13}$$

$$\boldsymbol{n}\times(\boldsymbol{H}^+ - \operatorname{grad}\Phi^-) = \boldsymbol{n}\times\boldsymbol{H}_e \tag{3.14}$$

where $\boldsymbol{H}_e$ expresses an external (exciting) field.

In this case eddy currents are calculated from the fundamental rule

$$\boldsymbol{j} = \operatorname{curl}\boldsymbol{H} \tag{3.15}$$

It should be stressed here that the problem formulated above has one great advantage, namely the field equations are considered separately inside and outside the conducting region. Both regions and also both functions $\boldsymbol{H}$ and Φ meet each other on the boundary, thus the formulation naturally belongs to the boundary methods, which will be shown in Section 4.3.

The use of the direct field quantities, although physically appropriate, poses some difficulties, especially if one considers the two-dimensional problems with the *current excitation*, as defined in Section 2.1. Then two components of vector $\boldsymbol{H}$ exist, so that two differential equations are to be solved which makes the process of computation twice as long. This situation leads directly to the concept of potential.

3.1.2 *Potentials*

From mathematical physics it is known that there exists a large number of potentials which are also somewhat arbitrary. Their examination is a huge task open to human creativeness. However, we limit ourselves the few, which are mostly used in the subject considered. The main purpose for introducing potentials is that they help in overcoming some difficulties mentioned above and related to the field variables. Simply speaking, the use of potentials enables us, in a simple way, to reduce the number of equations to be solved. The main drawback is that by using the potentials one escapes from physical reality into mathematical abstraction. However, they cannot be avoided in modern analysis that is based on mathematical modelling.

The excellent treatises on potentials used in applied electromagnetism are given by Carpenter (1977) and Hammond (1982) and the reader who wants to find both physical and mathematical considerations on them should resort to these works.

Generally speaking the potentials can be of scalar or vectorial nature. According to Helmholtz's[1] theorem each vector field is determined by two kinds of sources: curl sources and divergence sources. Considering a field vector $\boldsymbol{F}$ two extreme states may occur: solenoidal field (when $\operatorname{div}\boldsymbol{F} = 0$) and irrotational field (when $\operatorname{curl}\boldsymbol{F} = 0$). Thus, making use of the well-known vector identities one can represent the field vector in the following forms:

- irrotational field $\boldsymbol{F} = -\operatorname{grad} V$,
- solenoidal field $\boldsymbol{F} = \operatorname{curl}\boldsymbol{P}$.

Hence in the first case a scalar potential V is introduced while in the second one a vector potential $\boldsymbol{P}$ is used. If in a particular case both limit states do not occur, this case requires two potentials. Then the field vector is represented by the superposition of both vector and scalar potentials. In eddy current problems this is usually the case, that is, the field is solenoidal for both magnetic and current field ($\operatorname{div}\boldsymbol{B} = 0$ and $\operatorname{div}\boldsymbol{J} = 0$) so that vector potentials have to be considered.

In mathematical electromagnetism the above potential occurs in two basic forms: the first one is the magnetic vector potential and the second one the electric vector potential. In practice, since $\operatorname{div}\boldsymbol{B} = 0$ we have $\boldsymbol{B} = \operatorname{curl}\boldsymbol{A}$ ($\boldsymbol{A}$ is the magnetic vector potential) and since $\operatorname{div}\boldsymbol{J} = 0$ then $\boldsymbol{J} = \operatorname{curl}\boldsymbol{T}$ ($\boldsymbol{T}$ is the electric vector potential). Both potentials are useful in two-dimensional analysis because, by using them it is possible to transform a vectorial problem into a scalar one. As to the electric vector potential it is used in two-dimensional analysis as the so-called *stream function* u. Here we have one function instead of two components of a vector. It is well-known that the magnetic vector potential shows the same behaviour. This occurs in two-dimensional problems excited by infinite currents where the magnetic vector potential has only one component, namely that of the current direction.

Although we speak here of the vector potentials, the scalar potentials cannot be avoided since they are introduced by the curl-free fields or more formally by vector identities. Indeed, according to the definitions of vector potentials presented above, the vector potential defined is determined within the accuracy of the gradient of some scalar potential:

$$\boldsymbol{P} = \boldsymbol{P}' + \operatorname{grad} V \tag{3.16}$$

[1]Hermann Ludwig Ferdynand Helmholtz (1821–1894). German physicist, mathematician and physiologist working on the theory of differential equations.

while the scalar potential is within the accuracy of a constant C:

$$V = V' + C \tag{3.17}$$

The transformations $\boldsymbol{P} \mapsto \boldsymbol{P}'$ and $V \mapsto V'$ do not affect the field vector ($\boldsymbol{B}$ and $\boldsymbol{J}$ are invariant), which is the only physical quantity. The affected quantity is the scalar potential associated to a corresponding vector potential, that is, the magnetic vector potential $\boldsymbol{A}$ or the electric vector potential $\boldsymbol{T}$. The scalar potential representing an electric field part enables to consider the terminal conductor conditions or multiply connected conductor configurations (see Section 5.1). Dealing with the scalar potential is commonly recognized as one of the main problems in the eddy current analysis. In some papers it is referred to as the 'grad ϕ' problem. This problem gained more importance when the three-dimensional problems were encountered. Then the scalar fields are coupled to the vector ones inherently.

In two-dimensional cases the gradient of a scalar potential related to the vector potential $\boldsymbol{A}$ is constant and can be eliminated by transformation. If not, an additional component of the unknown vector may be introduced. In three-dimensional cases it is not so simple because the scalar potential is an additional vector. What is more, we have to ensure the uniqueness of the solution which strongly depends on the pair of potentials used. The problem is discussed in many papers and an excellent analysis has been carried out by Carpenter (1977) as well as by Trowbridge (1986). The paper of one of the authors (Krawczyk and Turowski 1988) also contains some relevant comments.

Let us now show the main properties of the potentials defined, indicating also their advantages and drawbacks. It does not mean, however, that other potentials which are not mentioned here are of lesser use. It may be that further activity in this field will bring up new potentials although the potentials given here are at present frequently and widely used.

Magnetic vector potential

The magnetic vector potential $\boldsymbol{A}$ is of great use and applicability when dealing with two-dimensional problems. Then the boundary conditions are expressed simply by the potential (that is, scalar here) itself or its normal derivative. Moreover, the boundary conditions expressed in terms of the magnetic vector potential are physically clear especially when considering an interior problem. Then Dirichlet's condition (if constant) means that the boundary is given by the highly conducting material and the Neumann condition (if homogeneous) indicates that the boundary represents a highly permeable material.

Introducing the definition of the magnetic vector potential ($\boldsymbol{B} = \operatorname{curl} \boldsymbol{A}$) into the fundamental equations (2.1–2.3) leads to the following set of equations:

- inside eddy current region

$$\operatorname{curl}\operatorname{curl} \boldsymbol{A}(\boldsymbol{x}, t) - \mu\gamma \frac{\partial \boldsymbol{A}}{\partial t} = \gamma \operatorname{grad} \phi \tag{3.18}$$

- outside eddy current region

$$\operatorname{curl}\operatorname{curl} \boldsymbol{A}(\boldsymbol{x}, t) = 0 \tag{3.19}$$

The above model requires the solution of a vector equation outside the eddy current region. To avoid this, one can introduce the second scalar potential Ω (now magnetic). Equation (3.18) is still valid for the eddy current region. The non-conducting region is now described by

$$\nabla^2 \Omega = 0 \tag{3.20}$$

where Ω results from $\boldsymbol{H} = -\operatorname{grad} \Omega$.

The latter model is called $A - \phi - \Omega$ while the previous one $A - \phi$.

As one can see, vector potential $\boldsymbol{A}$ is coupled with a scalar one which simply results from the definition of potential.

Eddy currents are calculated by using eqn (2.2) and in terms of potentials we have:

$$\boldsymbol{j} = -\gamma \frac{\partial \boldsymbol{A}}{\partial t} - \gamma \operatorname{grad} \phi \tag{3.21}$$

Then the condition of current continuity is as follows:

$$\operatorname{div} \gamma \left(\frac{\partial \boldsymbol{A}}{\partial t} - \gamma \operatorname{grad} \phi \right) = 0 \tag{3.22}$$

To obtain the values of eddy current densities one has to calculate both potentials by solving eqns (3.20) and (3.22). This seems to be a difficult task. Fortunately, one can simplify the problem applying the freedom in mutual relation between $\boldsymbol{A}$ and ϕ. It is known as *gauging* procedure.

In the theory of electromagnetism there are two gauge conditions which are commonly used:

- Coulomb's gauge

$$\operatorname{div} \boldsymbol{A} = 0 \tag{3.23}$$

- Lorentz's gauge[2]

$$\operatorname{div} \boldsymbol{A} = \mu\gamma\phi \tag{3.24}$$

[2] In fact, Lorentz's gauge has been introduced to wave problems and concerns also displacement current; here we consider, say, low frequency.

The first condition leads to the following set of equations describing the eddy current region:

$$\nabla^2 \boldsymbol{A} - \mu\gamma \frac{\partial \boldsymbol{A}}{\partial t} = \gamma \operatorname{grad} \phi \tag{3.25}$$

$$\nabla^2 \phi = 0 \tag{3.26}$$

while the Lorentz's gauge introduces the separation of potentials, so that we have:

$$\nabla^2 \boldsymbol{A} - \mu\gamma \frac{\partial \boldsymbol{A}}{\partial t} = 0 \tag{3.27}$$

$$\nabla^2 \phi - \mu\gamma \frac{\partial \phi}{\partial t} = 0 \tag{3.28}$$

In eddy current problems the Coulomb gauge is mostly used, so that the scalar electric potential plays an essential role in the computation process. The problem of how to adopt the magnetic vector potential by considering all above conditions to numerical practice has been analysed in depths by, among others, Biro (1988) and Song and Ida (1991).

If the model is two-dimensional the problem is slightly easier since the gradient of a scalar potential is constant in the region considered. The problem of determining the scalar potential for two-dimensional is called *the current constraint problem* because it consists in imposing some total condition on the flow of the total current. This condition depends on the way the long conductors are connected at infinity and on whether their circuits are passive or active. In general, this leads to the integro-differential equation since $\operatorname{grad} \phi$ is expressed as follows:

$$\operatorname{grad} \phi = \frac{1}{S} \int_S \frac{\partial A}{\partial t} dS + \frac{I}{S} \tag{3.29}$$

where I is the total current flowing through the cross-section S.

There are special cases (still two-dimensional) in which $\operatorname{grad} \phi$ is equal to zero unconditionally. Thus the analysis of the $\operatorname{grad} \phi$ problem is always necessary before starting the computational process. Indeed, there are many papers devoted to this problem and an extensive, even exhaustive, survey is given by Konrad (1985). Some researchers (Nakata *et al.* 1988) tried to investigate the physical meaning of the scalar potential considering pre-initial information on the phenomenon considered.

As for the three-dimensional case an opinion was expressed (Emson and Simkin 1983) indicating that the problem of $\operatorname{grad} \phi$ can be solved by introducing the modified vector potential $\boldsymbol{A}^*$. This potential results from the transformation which makes the problem (3.25)–(3.26) gradientless. Thus by letting

$$\boldsymbol{A}^* = \boldsymbol{A} + \frac{1}{\gamma}\int_0^t \operatorname{grad}\phi d\tau \tag{3.30}$$

eqn (3.25) takes the form

$$\nabla^2 \boldsymbol{A}^* = \mu\gamma \frac{\partial \boldsymbol{A}^*}{\partial t} \tag{3.31}$$

Note that in this case the gauge condition is

$$\operatorname{div} \gamma \boldsymbol{A}^* = 0 \tag{3.32}$$

It may be stressed here once more that for two-dimensional eddy current problems the magnetic vector potential is the most appropriate and is unmatched regarding field variables but in the case of three-dimensional modelling the benefits of its use are doubtful.

The commonly used gauge condition $\operatorname{div} \boldsymbol{A} = 0$ creates the possibility of introducing the second-order magnetic vector potential $\boldsymbol{W}$ defined as follows:

$$\boldsymbol{A} = \operatorname{curl} \boldsymbol{W} \tag{3.33}$$

The potential $\boldsymbol{W}$ can also be used in eddy current analysis but in very special cases, namely if a certain kind of symmetry exists (Hammond 1982). Some authors (Kriezis and Xypteras 1979) argue for this approach in spite of the fact that it lacks generality.

Electric vector potential

The potential $\boldsymbol{T}$ was introduced in the 1970s, and its theory and application is still developing and spreading. The main advantage of this vector is that it gives the possibility of joining dielectric and conducting regions. Furthermore, it leads directly to the determination of vector $\boldsymbol{H}$ since the relation between two vectors is linear. This vector potential is defined as

$$\boldsymbol{J} = \operatorname{curl} \boldsymbol{T} \tag{3.34}$$

and has as a consequence $\boldsymbol{H} = \boldsymbol{T} - \operatorname{grad} \Omega$, which leads to the following equations:

- inside eddy current region

$$\operatorname{curl}\operatorname{curl} \boldsymbol{T} = \mu\gamma\left(\frac{\partial \boldsymbol{T}}{\partial t} - \operatorname{grad}\Omega\right) \tag{3.35}$$

$$\nabla^2 \Omega = \operatorname{div} \boldsymbol{T} \tag{3.36}$$

- outside eddy current region

$$\nabla^2 \Omega = 0 \tag{3.37}$$

In this case the scalar potential Ω represents the magnetic field. The boundary conditions in interior problems (Dirichlet or Neumann) do not posess a physical meaning, so that they are difficult to prescribe. In exterior problems there are interface conditions which are also not entirely defined. Some authors strongly argue for the use of this vector for the reason that in special cases vector $\boldsymbol{T}$ is determined on the boundary of a conducting region ($\boldsymbol{T} = 0$) which is, of course, a very convenient feature in numerical solutions (Krakowski 1987). However, this feature is not general and concerns only some special geometrical configurations. Generally the normal component T_n is indeterminate so that the problem cannot be confined to the conducting region. The problem of continuity or discontinuity of vector $\boldsymbol{T}$ or, better say, its normal component T_n is the subject of many publications made in the 1980s. To conclude, one can say that scalar potentials are not tractable because of the difficulties arising from the boundary and interface conditions.

Stream function

For some special cases the electric vector potential $\boldsymbol{T}$ may be assumed to have one component only. Thus, instead of a vector quantity one considers a scalar one. Such a procedure transforms three-dimensional problem into two-dimensional problem. This scalar quantity, in other terms the *stream function*, is used for thin plates, that is, such plates whose thickness is small enough to neglect the variation of exciting field across their width. Moreover, taking one-component vector $\boldsymbol{T}$ implies that the exciting field has also one spatial component. The function, denoted here by u, is defined as follows

$$\boldsymbol{j} = 1_n \times \operatorname{grad}_t u \tag{3.38}$$

where 1_n is the unit vector in the direction normal to the surface of the plate and t is the direction tangential to it (Fig. 3.1).

Substituting function u to the fundamental equations leads to the integro-differential formulation with homogeneous Dirichlet boundary conditions ($u = 0$ on the boundary) which is as follows:

$$\nabla^2 u = \mu\gamma \left(\frac{\partial H_0}{\partial t} + \frac{\partial H'}{\partial t} \right) \tag{3.39}$$

where

$$H'(M) = \frac{1}{4\pi} \int_S K(M,N) u(N) dS \tag{3.40}$$

and $K(M, N)$ is the kernel of the integral and is expressed as

$$K(M,N) = \frac{\partial}{\partial n_M} \frac{\partial}{\partial n_N} \left(\frac{1}{r_{MN}} \right) \tag{3.41}$$

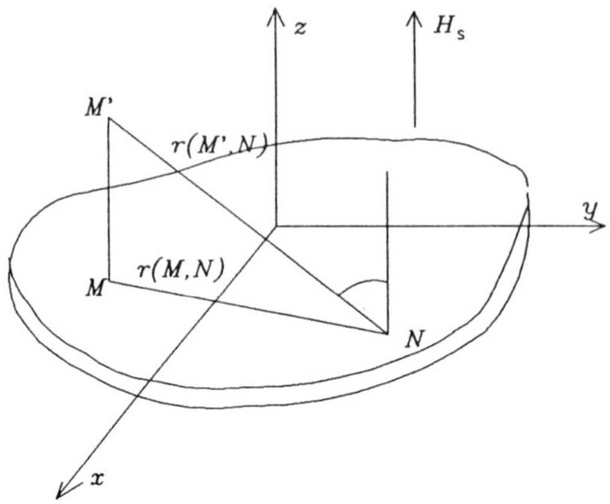

FIG. 3.3. The thin plate

H_0 represents the exciting field.

Furthermore, some authors (for example, Krakowski 1988) prove that the problem can be represented in the plain differential form, if special conditions are assigned concerning the frequency, the conductivity, or the thickness of a plate. In other words one can say that if the eddy current reaction is weak we have a pure differential equation of the parabolic type

$$\nabla^2 u - \mu\gamma \frac{\partial u}{\partial t} = \mu\gamma \frac{\partial H_0}{\partial t} \tag{3.42}$$

If the eddy current reaction is very weak (resistance-limited case) then Poisson's equation holds:

$$\nabla^2 u = \mu\gamma \frac{\partial H_0}{\partial t} \tag{3.43}$$

The latter case, however, seems to be oversimplified and is limited to a very narrow range of problems. On the other hand, the first approximation given by the parabolic equation (or Helmholtz's equation for the quasi-stationary case) is being used very often and mostly gives satisfactory results.

The potential u allows us to compute the eddy current density in a rather simple manner according to its definition (see eqn (3.38)). Indeed, assuming that the plate is flat and placed on the $0xy$ plane and the exciting field has a z-component, we have

$$J_x = \frac{\partial u}{\partial y} \tag{3.44}$$

$$J_y = -\frac{\partial u}{\partial x} \tag{3.45}$$

It seems that the potential described above can be very useful in many technical applications in which the assumption as to the small thickness of the plate is permissible.

In this section we have attempted to give some general hints on how to use and apply all the quantities describing a field. This is indeed a difficult task as one cannot strictly prove that this or that quantity is better than another one. As is always in the real world, the most appropriate method should be selected to fit each individual case. However, to do this we need to know all the features of each quantity.

3.2 Mathematical description

From the very beginning of electromagnetism scientists have used two ways of mathematically expressing electromagnetic phenomena. The first is the differential way while the second is the integral one. Both approaches are equivalent and interchangeable on the level of fundamental laws. For example, the Faraday's law which is of integral nature in space has differential form. This dualism of forms is not surprising since in the mathematical analysis there are theorems and formulas joining both forms. These are the Stokes, Gauss, and Green theorems which make this dualism mathematically clear and understandable. From a philosophical point of view there is a difference between the two forms, namely the differential one follows from a value of the field at a point to express the whole while the integral one starts to look at a system as a whole and goes to the field at a point. Following this consideration, the two methods of analysis appear to correspond to the inductive and deductive processes, respectively. This picture holds also when we try to establish a mathematical model of eddy current phenomena. A true dichotomy lies in front of us. In this case our choice has not only philosophical significance but is essential and in a way critical. The proper choice of the formulation makes our further activity either more fruitful or more 'in vain'. Simply speaking the formulation should be adequate for the problem in question. Thus the main formulations will be described here on the basis of the very fundamental division, namely integral or differential.

3.2.1 *Differential form*

As was mentioned earlier, some of the fundamental electromagnetic laws are formulated differentially. There is a controversy as to whether nature is differential or integral but one should note that the usage of differential equations is considerably wider than that of integral equations. Perhaps it stems from the fact that the mathematical background of differential equations is much older than the mathematics of integral equations. Roughly speaking, the first numerical results as a solution of integral equation was reached by Neumann in 1877, who used the iteration method (successive approximations). Needless to add that at that time solutions of differential equations existed widely.

The differential formulation contains the differential equation and boundary conditions, which can be both of space and time dependence. In this way the solution is unique as the boundary conditions confine the spectrum of solutions. From a philosophical point of view such an approach is a classical inductive one, that is, it goes from an elementary observation to the system as a whole. Our interest is focused on eddy current problems and so on Helmholtz's or parabolic equations for quasi-stationary or transient problems, respectively. Let us now show the classical differential formulation for the two-dimensional case expressed in terms of the magnetic vector potential.

Let the domain Ω be a cross-section of a conductor carrying a known time-dependent current and surrounded by ferromagnetic walls and on some part of the boundary by air. The total boundary is denoted by Γ. An example of this configuration is the slot of an electrical machine. Then the boundary conditions are as follows:
on the ferromagnetic walls (Γ_1)we have the homogeneous natural (Neumann) boundary condition while on the air part (Γ_2) the constant essential (Dirichlet) boundary condition.
The latter means that the field is assumed to coincide with this part of the conductor boundary. It was mentioned earlier that in this case the magnetic vector potential is the most appropriate and so is therefore being used here. Then we have (after elimination of grad ϕ)

- for the quasi-stationary case

$$\begin{aligned} \nabla^2 A(x,y) - i\omega\alpha^2 A &= -\mu j(x,y) \\ A_{\Gamma_2} &= f(x,y) \\ \frac{\partial A}{\partial n}\,|_{\Gamma_1} &= g(x,y) \end{aligned} \tag{3.46}$$

- for the transient case

$$\nabla^2 A(x,y,t) - \alpha^2 \frac{\partial A}{\partial t} = -\mu j(x,y,t) \tag{3.47}$$

  and the same boundary conditions. For both cases $\alpha^2 = \mu\gamma$.

Of course, the air–conductor interface can also be fitted by Neumann's condition which follows from Ampere's law. These two cases are interchangeable.

The differential form is the departure point for the finite difference method. Of course, this constitutes also the basis for other formulations, as was mentioned above, but these are treated separately and discussed later on.

3.2.2 *Integral form*

Let us now look at the eddy current problem from the integral point of view. This case is described by the integral equation and the integrally expressed constraint condition. Thus the philosophy of the approach is quite different from the previous one, namely, we use here the deductive way, that is, one proceeds from the total view of a problem to the elementary information. The key-point of integral formulation is Green's function and potential theory.[3] But before starting to use the sophisticated mathematical theory let us consider the problem in more heuristic way. Let current j flow through a filament placed in free space along, say, the z axis. Then the field problem is two-dimensional on the xy plane. Using Biot–Savart law for a filament one can determine the z component (only existing) of the magnetic vector potential A at a point P of space, as follows:

$$A_z(P) = \frac{1}{2\pi} j \ln \frac{1}{r_P} \tag{3.48}$$

where r_P is the distance between the conductor and the field point P. If the conductor has a finite cross-section Ω, then it can be considered to be composed of a number of filaments parallel to the z axis. It is obvious that the potential A_z may be calculated as the sum of all filament contributions, and this sum can be, of course, represented by an integral, so that, finally we have

$$A_z(P) = \frac{1}{2\pi} \int_\Omega j(M) \ln \frac{1}{r_{PM}} d\Omega_M \tag{3.49}$$

where r_{PM} is the distance between the source filament and the field point P, so that

$$r_{PM} = \sqrt{(x_P - x_M)^2 + (y_P - y_M)^2} \tag{3.50}$$

The same result can be reached using the theory of differential equations and the above integral is known there as the solution of Poisson's equation called Poisson's integral, and in the theory of potential the integral (3.49) is called the volume or Newtonian potential. Anyway, having the potential A determined, one can employ the fundamental electromagnetic relation and express the eddy current density in the conductor:

$$j(P) = -\gamma \operatorname{grad} \phi - \gamma \frac{\partial A}{\partial t} \tag{3.51}$$

where ϕ is a scalar potential resulting from the fact that the vector potential A is not unique (see Section 3.1).

[3] George Green (1793 – 1841), a miller from Nottingham, played an important role in the theory of electromagnetism and more generally in the theory of differential equations. His outstanding essay from 1824 was unknown until William Thomson (later Lord Kelvin) revealed it some decades later.

Considering the sinusoidal case with an angular frequency ω, eqn (3.51) takes the form

$$j(P) = -\gamma \operatorname{grad} \phi - i\gamma\omega A \tag{3.52}$$

Then making a simple substitution (3.49) to (3.52) one obtains the integral equation, which is recognized in mathematics as the Fredholm's equation of the second kind:

$$j(P) + \lambda \int_{\Omega} j(M) K(P, M) d\Omega_M = -\gamma \operatorname{grad} \phi \tag{3.53}$$

where $\lambda = i\omega\mu\gamma$.

It can be proved that in the two-dimensional case grad ϕ is constant in space and can be evaluated from the integral condition expressing the total current flowing through the conductor:

$$\int_{\Omega} j(P) d\Omega = I_t \tag{3.54}$$

Hence one needs the information on the total current in order to calculate the eddy current distribution.

But this problem is not analogous to the differential problem as formulated above. The reason is that this case discusses the conductor placed in free space without boundary conditions. If one wants to put some boundary restrictions then the kernel of the integral equation takes into account this fact.

We now approach in some way the notion of Green's function. This function contains two components: one is the fundamental solution G_0 which solves the following equation in free space:

$$\nabla^2 G_0(\boldsymbol{x}, \boldsymbol{x}') = \delta(r) \tag{3.55}$$

where $r = \mid \boldsymbol{x} - \boldsymbol{x}' \mid$ and δ is the Dirac or impulse function defined as follows:

$$\delta(r) = \begin{cases} 1 & \text{for } r = 0 \\ 0 & \text{elsewhere} \end{cases}$$

and

$$\int_{-\infty}^{+\infty} \delta(r) dr = 1$$

while the second part is such a solution of the Laplace's equation:

$$\nabla^2 F = 0 \tag{3.56}$$

which enforces boundary conditions to be satisfied.

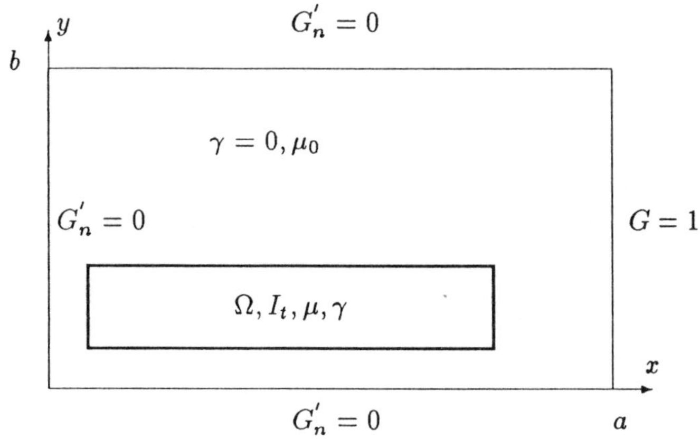

FIG. 3.4. The arrangement of the problem

As an example we shall present the solution of Green's function for the case when one looks for the distribution of eddy currents in the conductor placed in a closed domain having strictly prescribed boundary conditions (Fig. 3.4). In this case Green's function can be found in a closed form as the solution of the boundary value problem (Lipinski and Krason 1981):

$$\nabla^2 G(\boldsymbol{x}, \boldsymbol{x}') = \delta(r) \tag{3.57}$$

Solving eqn (3.57) with the boundary conditions shown in Fig. 3.4 one obtains Green's function expressed as a Fourier series, so that the complete Green's function is as follows:

$$G = \sum_{n=1,3,\ldots}^{\infty} C_n f_n(x) g_n(y) \tag{3.58}$$

where

$$f_n(x) = \cosh(\lambda_n x)$$
$$g_n(y) = \cos(\lambda_n y)$$
$$C_n = -\frac{4}{n\pi \cosh(\lambda_n a)}$$

Then the integral equation for the problem is:

$$j(M) = C + i\omega\mu\gamma \int_{\Omega} j(P) G(P, M) d\Omega_P \tag{3.59}$$

The unknown constant C, representing the electric field (see eqn (3.53)), results from the integral condition which relates the eddy current distribution with the known total current flowing through the conductor. Indeed, one can write:

$$\int_{\Omega} j(M)d\Omega = I_t \tag{3.60}$$

It should be stressed that the integral formulation is not as easy for the interior eddy current problem as it is for the external problem. This results from the fact that the kernel of the integral equation is more complex when considering boundary conditions. On the other hand, the case described here as the differential example is rather rare because usually we deal with the so-called 'open boundary problems' or with problems in which the eddy current region is much smaller than the whole space. In the latter the integral formulation also has an advantage, namely it allows deleting a large non-conducting area from our consideration. For these cases the integral formulation is most appropriate.

The integral formulation, based on volume integrals, is naturally the basis for adopting the method of finite sums or the collocation method which will be presented in Chapter 4.

3.2.3 *Variational form*

It is well-known that the majority of differential problems also have the variational property and *vice versa.* This fact is known even from the Euler–Lagrange equations which differentially describe Hamilton's principle.[4] This is valid as long as conservative systems are being considered. It cannot be hidden that eddy current problems are of different nature, namely they create a dissipative system. And so the question arises: 'should we consider the variational forms to be outside the scope of our considerations as inappropriate?'. The answer is that they are really useful.

The former chapters showed that as far as we deal with sinusoidal cases we can use the Fourier transformation, which leads to the Helmholtz equation instead of the parabolic one and it can be treated as a conservative system. Indeed, the transformation mentioned above makes a 'snapshot' of the phenomenon at one time instant and at this instant all processes are not dissipative. The non-sinusoidal (transient) case is far more complex regarding this point of view, but there is also some possibility of applying variational-like formulation. Indeed, the variational approach is close to both the differential one and the integral one. Because of the fact that the variational approach constitutes a problem of optimization, it needs some restrictions and these are the boundary conditions. This feature makes the variational approach closer to the differential formulation. On the other

[4] W.R. Hamilton (1805–1865)—Irish mathematician and astronomer working on mechanics (the principle quoted) and mathematics (quaternions)

hand, the variational formulation takes, in the nature of things, integral form. Philosophically speaking, the variational approach describes the object as a whole, excluding the boundary which is observed as a set of points.

The variational approach can be derived in various ways, namely from the Euler–Lagrange equation or more physically from energetic considerations. Let us now proceed with the mathematical background of this formulation.

Theorem 3.1 *Let the operator L be self-adjoint and positively determined. Then the function u solving the operator equation*

$$Lu = f$$

minimizes the following functional $J(u)$

$$J(u) = (Lu, u) - 2(f, u)$$

The reader can find proof of the theorem in many books (for example, Finlayson 1972), so that it seems to be unnecessary to repeat it here. Going to the most commonly used form one has to employ now the first Green identity, which is:

$$(Lu, u) = (L_{-1}u, L_{-1}u) + B(u, u) \tag{3.61}$$

where (a, b) denotes scalar product of a and b[5] in the domain of consideration, L_{-1} denotes the operator which is of reduced order in functional space with respect to the operator L, and B is the boundary functional. After these very general remarks we pass into eddy currents. Then the operator L for quasi-stationary harmonic problems is as follows:

$$L \equiv \nabla^2 - \alpha^2 \tag{3.62}$$

where ∇^2 is the Laplace's operator and $\alpha^2 = \mu\gamma$.

It can be proved that this operator is self-adjoint and positively determined, so that the above theorem is valid. In this case the operator L_{-1} is the gradient operator and the functional B is expressed as follows:

$$B(u, u) = \int_\Gamma u \frac{\partial u}{\partial n} d\Gamma \tag{3.63}$$

Hence the aim of the variational problem is: *to find the function u which minimizes the functional*

$$K(u) = (\operatorname{grad} u, \operatorname{grad} u) - \alpha^2(u, u) - 2(f, u) - \left[u, \frac{\partial u}{\partial n}\right] \tag{3.64}$$

[5] in the sense of Hilbert's space, $(u, v) = \int uvdS$

where f is the function of excitation and $[a,b]$ denotes the scalar product of a and b acting on the boundary.

The problem can be solved by many methods even analytically (Sikora *et al.* 1978). Some of them have been employed in the finite element, method which will be described in detail in Chapter 4. One should stress here the fact that the variational formulation requires the boundary conditions to be prescribed since the boundary integral occurs in the expression (3.64). Hence it can be used in the analysis of problems involving closed boundaries. Let us underline once again that this requirement relates the variational form to the differential one in spite of its integral nature.

Very close to the variational formulation is the so-called 'weak formulation'. As far as the elliptic problem is concerned a complete correspondence exists between the two formulations, since the operator is positively determined and self-adjoint. But this is not the case for the parabolic problem which mathematically describes eddy currents in the transient state. In this case the problem is dissipative by all means and it is basically difficult to formulate the variational principle. Nevertheless, some authors attempted to do it (Gurtin 1964; Tarasiewicz 1987). However, those attempts were of mathematical nature rather than using physical considerations. That is why we shall introduce here the weak formulation, which gives the possibility of treating dissipative problems in a quasi-variational way. The procedure is as follows:

Let us take the parabolic equation:

$$\nabla^2 u(\boldsymbol{x},t) = \alpha^2 \frac{\partial u}{\partial t} - f(\boldsymbol{x},t) \tag{3.65}$$

Multiplying both sides of the equation by the arbitrary function v and integrating both sides over the spatial domain Ω one obtains the following integral equation:

$$\frac{d}{dt}\alpha^2(u,v) + a(u,v) = (f,v) + [u,v] \tag{3.66}$$

Then the problem becomes of integral nature but with respect to time it still remains differential. There are various ways of tackling time- derivative problems. Here one can adopt the well-known linear one-step method, that is, the so-called θ-method which is the simplest one. This approximates function u by a piecewise linear function and for each piece one can write:

$$u = u^n\theta + u^{n+1}(1-\theta) \tag{3.67}$$

In this approximation the time derivative is

$$\frac{du}{dt} = \frac{u^{n+1} - u^n}{\Delta t} \tag{3.68}$$

Then eqn (3.66) becomes

$$(u^{n+1}, v) + a(\theta u^{n+1}, v) = (u^n, v) - a((1-\theta)u^n, v)$$
$$+(f^{n+1}\theta + f^n(1-\theta), v) + [v, (\frac{\partial u}{\partial n})^{n+1}\theta] + [v, (\frac{\partial u}{\partial n})^n(1-\theta)] \quad (3.69)$$

As can seen, this procedure posseses a recurrent nature since starting from the initial condition one proceeds step-by-step, using at each step the information obtained at the previous one. This formulation may be considered as the starting point for other numerical methods, especially the finite element method. Similarly we can set up the weak formulation for the wave propagation problem but this case is beyond the scope of this book.

There is a common opinion that all variational as well as weak formulations are suitable for employing the finite element method. Indeed, it is so, although there are no essential obstacles to use others such as the finite difference or collocation method. However, to be in accordance with the tradition we have to confirm the opinion mentioned and so we shall consider these formulations as a basis for the finite element approach.

3.2.4 *Boundary-integral form*

The variational approach described in the last chapter is mostly linked with the first Green identity while the boundary-integral approach is always associated with the second Green identity.

The second Green theorem or identity for the operator L is as follows:

$$(Lu, v) - (L^*v, u) = B(u, v) \quad (3.70)$$

where L^* is the adjoint operator to the operator L and v is an arbitrary function belonging to the functional space that is proper for u. The expression $B(u, v)$ is the boundary functional. Considering the equation

$$\nabla^2 u(\boldsymbol{x}) - \alpha^2 u(\boldsymbol{x}) = f(\boldsymbol{x}) \quad (3.71)$$

with boundary conditions prescribed one can write the exact second Green identity as follows:

$$(\nabla^2 u - \alpha^2 u, v) - (\nabla^2 v - \alpha^2 v, u) = \left[u, \frac{\partial v}{\partial n}\right] - \left[v, \frac{\partial u}{\partial n}\right] \quad (3.72)$$

The final shape of this formulation depends on function v and this function can be the fundamental solution of the problem considered or, in other words, the free-space Green function. To find function v we have to solve the following equation:

$$\nabla^2 v - \alpha^2 v = \delta(r) \quad (3.73)$$

The variable r is the distance between the source and field points. Equation (3.73) has the physical interpretation, namely one looks for the potential v caused by the current source placed in free-space.

The solution of eqn (3.73) is known and is as follows:

- for the two-dimensional problem

$$v(r) = \frac{K_0(\alpha r)}{2\pi} \tag{3.74}$$

- for the three-dimensional problem

$$v(r) = \frac{e^{-ikr}}{4\pi} \tag{3.75}$$

where K_0 is the modified Bessel function of zero order.

Let us consider the two-dimensional case only since the three-dimensional case is of vectorial nature and will be discussed separately. Thus the normal derivative of v is as follows:

$$\frac{\partial v}{\partial n} = \frac{\alpha}{2\pi} K_1(\alpha r) \frac{\partial r}{\partial n} \tag{3.76}$$

where K_1 is the modified Bessel function of unity order.

Using the property

$$(\delta(\mid \boldsymbol{x}_i - \boldsymbol{x} \mid), u(\boldsymbol{x})) = u(\boldsymbol{x}_i) \tag{3.77}$$

and substituting eqn (3.71) to eqn (3.72) after some mathematical rearrangements one can reach the final form leading to the function u at any arbitrary point $\boldsymbol{x}_i$ of the region bounded by the fixed boundary:

$$u(\boldsymbol{x}_i) = \left[v, \frac{\partial u}{\partial n} \right] - \left[u, \frac{\partial v}{\partial n} \right] + (f, v) \tag{3.78}$$

It should be noted that the boundary value problem considered cannot be given as a Cauchy problem, that is, at any point of the boundary the Dirichlet and Neumann conditions cannot be simultaneously prescribed. Thus eqn (3.78) is a formula only and does not allow solving the problem. That is why one should transform eqn (3.78) into a boundary-integral equation and this can be done by moving the internal point $\boldsymbol{x}_i$ onto the boundary. Then, the region of consideration will be common to all variables included in eqn (3.78). But in doing so one faces a singularity on the boundary, fortunately of a weak form. This means that all integrals are integrable in Cauchy's sense. Dealing with such integrals we can use the formulae known from a mathematical analysis or we can solve this problem

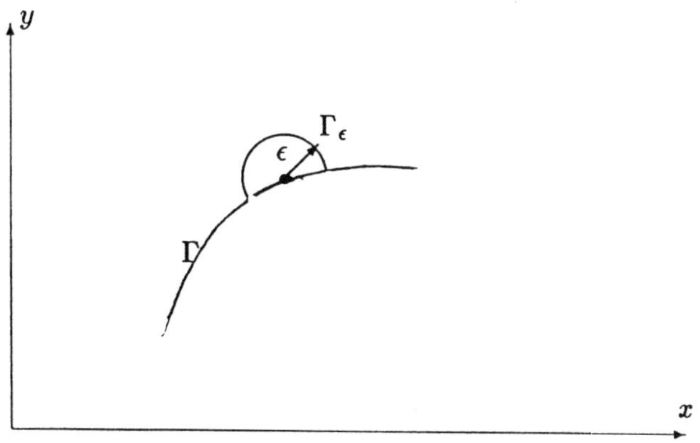

FIG. 3.5. The explanation of the singularity analysis

with some heuristic considerations. This method is used by the majority of those who created the boundary integral world and is discussed below.

Consider the smooth boundary and choose a singular point on it. To avoid this point we surround it by a semicircle or semisphere, respectively in two-dimensional and three-dimensional cases (Fig. 3.5). Radius ϵ is small enough to neglect the changes of function on this newly created part of the boundary. Then the integrals appearing in (3.78) can be presented as a sum of two integrals: one over the whole boundary except Γ_ϵ and the second just over this part. For the two-dimensional case the integrals to be established are as follows:

$$I' = \int_{\Gamma_\epsilon} \frac{\partial u}{\partial n} K_0(\alpha\epsilon) d\Gamma \tag{3.79}$$

$$I'' = \int_{\Gamma_\epsilon} u K_1(\alpha\epsilon) d\Gamma \tag{3.80}$$

Considering that ϵ tends to zero one obtains in both cases as a result of the second integral the value

$$I'' = \frac{2\pi - \eta}{2\pi} u(\boldsymbol{x})$$

while the first integral tends to zero.

Hence the final form of the boundary integral equation is as follows:

$$\frac{\eta}{2\pi} u(\boldsymbol{x}_i) + \left[u, \frac{\partial v}{\partial n}\right] = \left[\frac{\partial u}{\partial n}, v\right] + (f, v) \tag{3.81}$$

It is obvious that the angle coefficient will be equal to 0.5 if the point $\boldsymbol{x}_i$ lies on the smooth-enough part of the boundary (η equals π). In any other case it should be calculated.

Although the solution of this equation can be obtained by a few methods it is commonly recognized as a departure point to the boundary element method. Note that for pure boundary value problems, that is, Dirichlet or Neumann, the equation discussed becomes the well-known integral Fredholm equation of the first or second kind, respectively.

Let us now pass to the transient problem. Is there any Green theorem valid for this problem? In fact there is not, but one can generalize Green's formulae even for parabolic equations.

The boundary-integral formulation of the parabolic problem can be derived in three ways. These are:

(1) some type of the finite difference schemes;
(2) Laplace's transformation;
(3) the introduction of time-dependent fundamental solution.

Only the latter became widely used, so we shall describe it in more detail.

Before we go ahead with it we shall describe very briefly the remaining two. Coming back to the parabolic equation (3.65) one can apply to it the conventional finite difference scheme, similar to that used in the variational time formulation. Namely, we introduce the linear one-step method, so that the function u is represented by the following difference scheme with parameter θ:

$$u = u^{n+1}\theta + (1-\theta)u^n \tag{3.82}$$

where $\theta \in [0,1]$ and

$$\frac{\partial u}{\partial t} = \frac{u^{n+1} - u^n}{\Delta t} \tag{3.83}$$

Introducing the above into eqn (3.65) one obtains the Helmholtz equation

$$\theta\nabla^2 u^{n+1} - \Delta t \alpha^2 u^{n+1} = -(1-\theta)\nabla^2 u^n - \Delta t \alpha^2 u^n - f^{n+\theta} \tag{3.84}$$

The boundary-integral formulation of eqn (3.84) has been shown earlier. The small usefulness of this approach follows from the well-known problems with convergence and stability of the difference scheme.

The second way, consisting in introducing Laplace's transform, leads also to the Helmholtz's equation

$$\nabla^2\overline{u}(\mathbf{x},s) - \alpha^2 s\overline{u}(\mathbf{x},s) = u(\mathbf{x},0) + \overline{f}(\mathbf{x},s) \tag{3.85}$$

since we have

$$\mathcal{L}\left(\frac{\partial u}{\partial t}\right) = s\overline{u} - u(\boldsymbol{x},0) \tag{3.86}$$

But in this case the solution of integral equation depends on the parameter s and in order to obtain the solution in time space one has to make the inverse transformation. As the solution in transformation space is given in a discrete manner it is necessary to choose properly the sequence $\{s_i\}$. The inappropriate choice can give a time solution in the domain that will be unwanted or totally without interest. Moreover, the numerical inversion of the Laplace's transform is one of the classical examples of the ill-posed problems in Hadamard's sense. Thus the third way remains as the best and we shall focus our attention on it.

Let eqn (3.65) be again the basis of our considerations. Introduce the time-spatial domain in which we shall integrate both sides of the equation and multiply them by an arbitrary function v that depends on both time and space (John 1978). Now, having the integral identity we shall use the generalized Green formula which gives:

$$\int_0^t \int_\Omega (Lu - L^* v) d\Omega d\tau = \int_0^t \int_\Gamma \left(v \frac{\partial u}{\partial n} - u \frac{\partial v}{\partial n} \right) d\Gamma d\tau + \alpha^2 \int_\Omega uv d\Omega \,\Big|_{\tau=0}^{\tau=t} \tag{3.87}$$

Thus we arrived at the same point as before, that is, we obtained the integral formula that relates the searched function u and the arbitrary function v. Proceeding as before *mutatis mutandis* we have to find function v as a solution of the following equation:

$$L^* v = 0 \tag{3.88}$$

Note that in this case the operator L is not self-adjoint and it is expressed as

$$L^* \equiv \nabla^2 + \alpha^2 \frac{\partial}{\partial t} \tag{3.89}$$

The solution of eqn (3.88) is known as a thermal potential and its form is as follows:

$$v = \frac{\alpha^2}{[4\pi(t-\tau)]^{d/2}} \exp\left[-\frac{\alpha^2 r^2}{4(t-\tau)} \right] \tag{3.90}$$

where r is the distance between the point of observation and the source point, and d is the dimension of the problem considered.

Putting the above into eqn (3.87) and using the fact that

$$\lim_{\tau \to t} v = \delta(r) \tag{3.91}$$

one obtains the final formula for calculating the function u inside the domain in question at time t and at the point $\boldsymbol{x}_i$:

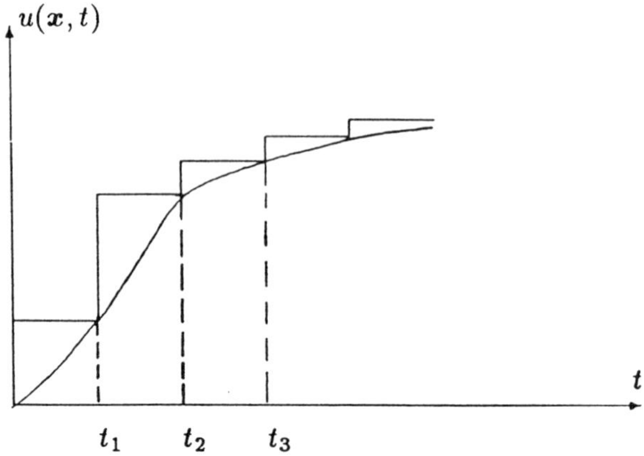

FIG. 3.6. Approximation in time

$$\alpha^2 u(\boldsymbol{x}_i, t) = \int_0^t \int_\Gamma (v\frac{\partial u}{\partial n} - u\frac{\partial v}{\partial n}) d\Gamma d\tau$$
$$+ \alpha^2 \int_\Omega uv d\Omega \mid_{\tau=0} + \int_0^t \int_\Omega fv d\Omega d\tau \tag{3.92}$$

The above expression is a formula only and does not allow calculating anything since, as it was already mentioned, simultaneous existence of Neumann and Dirichlet boundary conditions is impossible. Thus, in order to obtain a well-posed equation we have to move the internal point $\boldsymbol{x}_i$ onto the boundary.

Before we move the point onto the boundary let us make the first approximation of the problem in order to eliminate the time integrals. Suppose that both function u and its normal derivative can be replaced by the stair-case function in time (Fig. 3.6).

$$u(\boldsymbol{x}, t) = \sum_{i=1}^{m} u(\boldsymbol{x}, t_i)[\eta(t_i) - \eta(t_{i-1})] \tag{3.93}$$

where $\eta(t)$ is the unit step function and $t_m = t$. Then the time integrals are represented by sums. Changing the sequence of integration one obtains as spatial integrands such integrals which can be solved analytically both in the three-dimensional and two-dimensional case. Let us leave, however, the three-dimensional solution as it is vectorial requiring a separate description, and concentrate our attention on the two-dimensional one.

The analytical expressions of both time integrals are as follows:

$$\beta(\boldsymbol{x}; \boldsymbol{x}_i, t; t_i; t_{i-1}) = \int_{t_{i-1}}^{t_i} \frac{\partial v}{\partial n} dt = [\exp(w_{i-1}) - \exp(w_i)] \tag{3.94}$$

$$\varphi(\boldsymbol{x};\boldsymbol{x}_i,t;t_i;t_{i-1}) = \int_{t_{i-1}}^{t_i} v dt = \frac{\alpha^4 d}{8\pi^2 r}[E_1(w_i) - E_1(w_{i-1})] \quad (3.95)$$

where E_1 is the integral exponent

$$E_1(x) = \int_0^x \frac{e^{-t}}{t} dt$$
$$w_n = \frac{\alpha^2 r^2}{4(t-t_n)}$$

$r = |\boldsymbol{x} - \boldsymbol{x}_i|$, d is the distance between the point $\boldsymbol{x}_i$ and the integration point.

Thus, finally eqn (3.92) becomes:

$$\begin{aligned}\alpha^2 u(\boldsymbol{x}_i,t) + \sum_{i=1}^{m} \int_\Gamma u(\boldsymbol{x},t_i)\beta d\Gamma &= \sum_{i=1}^{m} \int_\Gamma \frac{\partial u}{\partial n} |_{t_i} \varphi d\Gamma \\ &+ \alpha^2 \int_\Omega v |_{\tau=0} u(\boldsymbol{x},0) d\Omega \\ &+ \int_0^t \int_\Omega f v d\Omega \end{aligned} \quad (3.96)$$

At this time we can move the internal point onto the boundary using the same technique as before concerning the singularity. As expected, the result is the same, that is, the integral over the semi-circle gives the contribution that depends on the smoothness of the boundary, and so the boundary integral equation becomes:

$$\begin{aligned}\frac{\eta}{2\pi}\alpha^2 u(\boldsymbol{x}_i,t) + \sum_{i=1}^{m} \int_\Gamma u(\boldsymbol{x},t_i)\beta d\Gamma &= \sum_{i=1}^{m} \int_\Gamma \frac{\partial u}{\partial n} |_{t_i} \varphi d\Gamma \\ &+ \alpha^2 \int_\Omega v |_{\tau=0} u(\boldsymbol{x},0) d\Omega \\ &+ \int_0^t \int_\Omega f v d\Omega \end{aligned} \quad (3.97)$$

Since the boundary integral formulation is given only in spatial terms it is obvious that this formulation is again the starting point of the boundary element method. Of course, the time approximation adopted here can be different, for example it may be of linear or higher order but this does not change the essence of the above derivations. That is why we have used here the simplest time approximation.

Passing to the three-dimensional analysis one has to underline the fact that the boundary-integral approach may employ a number of variables (see Section 3.1) but using direct variables (field variables) seems to be appropriate. Also the boundary-integral formulation itself can be established in various manners, for example, using vectorial Green formula (Stratton 1943, Mitra 1973). It seems, however, that for the purpose of eddy currents the way presented below is again the most appropriate. Indeed, considering an externally excited conducting solid, one can write the differential equation following directly fundamental electromagnetic equations:

$$\nabla^2 \boldsymbol{H}^+(\boldsymbol{x}, t) = \mu\gamma \frac{\partial \boldsymbol{H}^+}{\partial t} \tag{3.98}$$

$$\nabla^2 \phi^-(\boldsymbol{x}, t) = 0 \tag{3.99}$$

The above equations are coupled by the interface conditions:

$$\boldsymbol{n} \times (\boldsymbol{H}^+ - \operatorname{grad}\phi^-) = \boldsymbol{n} \times \boldsymbol{H}_e \tag{3.100}$$

$$\boldsymbol{n} \cdot (\mu\boldsymbol{H}^+ - \mu_0 \operatorname{grad}\phi^-) = \mu_0 \boldsymbol{n} \cdot \boldsymbol{H}_e \tag{3.101}$$

where $\boldsymbol{H}_e$ is an exciting field.

To formulate the boundary integral approach the concept of secondary sources will be employed (Tozoni and Mayergoyz 1975, Mayergoyz 1986). These sources are the surface current density and the surface magnetic charge distributed on the boundary. Thus, introducing the surface variables and using the potential theory theorems one obtains

$$\boldsymbol{H}^+(P, t) = \operatorname{curl} \int_S \boldsymbol{j}_s(M, t) K(r_{PM}, t, \tau) dS_M d\tau \tag{3.102}$$

where

$$K(P, M, t, \tau) = \frac{\alpha}{8\pi^{3/2}(t-\tau)^{3/2}} \exp(-\frac{\alpha^2 r^2}{4(t-\tau)}) \tag{3.103}$$

$$\phi^-(P, t) = \frac{1}{4\pi} \int_S \sigma(M, t) r_{MP}^{-1} dS \tag{3.104}$$

Introducing a staircase function spanned on n time intervals for the approximation of unknown variables $\boldsymbol{j}_s$ and σ and linking potentials by interface conditions one obtains the system of boundary-integral equations in the form of:

$$0.5\alpha^2 \boldsymbol{j}_s(P, t) + \int_S \boldsymbol{n}_P \times \sum_{i=1}^{n} \boldsymbol{j}_s(t_i) \times \boldsymbol{r}_{MP} \Psi(r_{MP}, t, t_i, t_{i-1}) dS$$
$$- \frac{1}{4\pi} \int_S \sigma(M, t_i) \boldsymbol{n}_p \times \boldsymbol{r}_{MP} \frac{1}{r_{MP}^3} dS$$

$$= -\boldsymbol{n}_P \times \boldsymbol{H}_e(t) \tag{3.105}$$

$$0.5\sigma(P,t) + \frac{1}{4\pi}\int_S \sigma(M,t)\boldsymbol{n}_P \cdot \boldsymbol{r}_{MP}\frac{1}{r_{MP}^3}dS$$
$$- \int_S \boldsymbol{n}_P \cdot \sum_{i=1}^{n} \boldsymbol{j}_s(M,t_i) \times \boldsymbol{r}_{MP}\Psi(\boldsymbol{r}_{MP},t,t_i,t_{i-1})dS$$
$$= -\boldsymbol{n}_P \cdot \boldsymbol{H}_e(t) \tag{3.106}$$

$$\Psi(r_{MP},t,t_i,t_{i-1}) = \frac{\alpha^2}{2\pi^{3/2}r_{MP}^3}[\Gamma(\tfrac{3}{2},w_{i-1}) - \Gamma(\tfrac{3}{2},w_i)] \tag{3.107}$$

where

$$w_i = \frac{\alpha^2 r_{MP}^2}{4(t-t_i)}$$

and $\Gamma(p,x)$ is the incomplete Gamma function and in this case is as follows:

$$\Gamma(3/2,w_i) = \sqrt{w_i}\exp(-w_i) + \frac{\sqrt{\pi}}{2}\mathrm{erf}(\sqrt{w_i})$$

Note that $\Gamma(3/2,0) = 0$ and $\Gamma(3/2,\infty) = \sqrt{\pi}/2$. For $t = t_n$ function Ψ can be expressed in simpler manner, namely

$$\Psi = \frac{\alpha^2}{2\pi^{3/2}r_{MP}^3}\left[\frac{\sqrt{\pi}}{2} - \Gamma(\tfrac{3}{2},w_{n-1})\right] \tag{3.108}$$

This leads to the important observation that for time tending to infinity function Ψ takes the form

$$\lim_{t\to\infty}\Psi = \frac{\alpha^2}{4\pi r^3} \tag{3.109}$$

Since the expression (3.109) represents the kernel of a double-layer potential in the static problems, it confirms the correctness of the above considerations.

For quasi-stationary, that is, sinusoidal problems the procedure is similar. The difference is in the kernel of eqn (3.103) since we have then:

$$K(r) = \frac{\exp(-ikr)}{4\pi r} \tag{3.110}$$

where $k = \sqrt{\omega\mu\gamma/2}$.

Consequently all subsequent kernels will change and, of course, the time considerations are no longer valid.

It should be stressed again that the use of the direct variables is not the only possibility of the three-dimensional boundary integral formulation.

There exists a number of papers showing the boundary-integral method in other contexts, for example, in T-Ω or A-ϕ formulations (Antonopoulos *et al.* 1992; Davey and Barnes 1983).

The boundary-integral approach is elegant as to its mathematics, but contains difficulties connected with singularities and rather cumbersome numerical integration. Nevertheless, the use of this approach increases year by year and attracts more and more followers.

4

METHODS OF NUMERICAL ANALYSIS

4.1 General remarks

Although the continuous mathematical model gives some essential and meaningful information on the problem considered, what we really need is a numerical result. The qualitative analysis is of limited use if it is not supported by numerical results, and therefore we need to discretize a worked-out continuous model. This procedure creates new knowledge which may be easily undestood by a mathematician but may be too difficult for an engineer. Thus, it is important to find an adequate compromise, and at this point we should warn the readers that they will not find here any mathematical proofs and considerations on the stability or convergence of the methods described. What we really want to convey to readers are some hints as to general features of the methods and how they can be applied to eddy current problems.

THe search for an approximated solution is not necessarily connected with the numerical methods. Analytical solutions are also loaded with truncation or approximation errors, since they are usually given in the series form. Thus one needs to approximate the infinite series by a finite number of terms. Also the iterative method gives approximated solutions but its nature is analytical. This kind of solution will be excluded from our consideration and in this book we shall focus our attention on the 'pure' numerical methods only.

Let us now define what the numerical method means here. For our purpose we call *numerical methods those procedures which transform the continuous mathematical model into a set of algebraic equations and this transformation is done by the discretization of the time-space considered.* The way of discretization determines the corresponding method and a name is assigned to it. Although some variants of each method are possible we shall try to describe all methods in their clear, widely used and commonly recognized form. Consequently, this part of the book constitutes its 'core' and therefore we shall devote to it far more attention than was done to previous chapters. These methods of approximation may be classified as to the magnitude which is to be approximated. Also one can classify the numerical methods as to the place where approximation is done and then one can distinguish 'domain methods' and 'boundary methods'. This, however, is decided upon when formulating the problem (see Section 3.2) and thus there is no choice when considering numerical approaches.

Having already fixed the continuous model that has a form of an operator equation, one can approximate the operator itself or the function. This fact divides all numerical methods into two families. We shall present these groups of methods by giving representative examples of methods. It is well-known that the approximation of the operator in its fundamental sense is mathematically less complicated than the approximation of the function. The latter one requires the analyst to be acquainted with the theory of approximation while the first one has very simple mathematical or even physical interpretation. Of course, the above remarks are valid for the most simple variant of both families. However, modern analysis can be more complex even for the methods with an approximated operator. Nevertheless, it is intuitively felt that the methods with an approximated operator have to be used in geometrically more simple cases, while the methods with an approximated function should be used in more complicated ones.

4.2 Approximation of operator

4.2.1 *Finite difference method*

The finite difference method has been well-known many years and continues to be popular. The main reason for its great usefulness is its very clear and simple interpretation, both mathematically and physically. The latter is, roughly speaking, the electric circuital interpretation while the first one results from the theory of Taylor's series expansion. This last aspect will also serve in describing the method.

Let us take the operator equation:

$$Lu = f \tag{4.1}$$

where L is the differential operator, u is the unknown function and f is the excitation or force function. It is obvious that this operator can be approximated ($L \to L_h$) with the parameter h. Generally, the operator contains the derivatives of the first and second orders as well as identity components. Using the Taylor expansion one can write:

$$u(\boldsymbol{x}_i + h) = u(\boldsymbol{x}_i) + hu^{'}(\boldsymbol{x}_i) + h^2u^{''}(\boldsymbol{x}_i) + O(h^3) \tag{4.2}$$

Neglecting all the components of order $h \geq 3$ one may write the formulae for the derivatives of the first and second order:

$$u^{'}(\boldsymbol{x}_i) = \frac{u(\boldsymbol{x}_i + h) - u(\boldsymbol{x}_i)}{h} \tag{4.3}$$

$$u^{''}(\boldsymbol{x}_i) = \frac{u(\boldsymbol{x}_i + h) - 2u(\boldsymbol{x}_i) + u(\boldsymbol{x}_i - h)}{h^2} \tag{4.4}$$

Substituting eqn (4.4) into eqn (4.1) and assuming the operator L as the one-dimensional Helmholtz's operator

$$L = \frac{\partial^2}{\partial x^2} - \alpha^2 \tag{4.5}$$

one obtains:

$$\frac{u(x_i + h) - 2u(x_i) + u(x_i - h)}{h^2} - \alpha^2 u(x_i) = f(x_i) \tag{4.6}$$

$i = 1 \ldots n$.

In the two-dimensional case the operator L is given as

$$L = \frac{\partial^2}{\partial x^2} + \frac{\partial^2}{\partial y^2} - \alpha^2$$

so that the same scheme is used in both directions: x and y. Then we have the so-called 'five-point star' and for the increments of mesh in orthogonal directions h_x and h_y we have

$$\begin{aligned} &\frac{u(x_i - h_x, y_i) - 2u(x_i, y_i) + u(x_i + h_x, y_i)}{h_x^2} \\ &+ \frac{u(x_i, y_j - h_y) - 2u(x_i, y_j) + u(x_i, y_j + h_y)}{h_y^2} - \alpha^2 u(x_i, y_j) = f(x_i, y_j) \end{aligned} \tag{4.7}$$

Equations (4.6) and (4.7) represent a system of equations, the solution of which are the values of function u. In the two-dimensional case when the function u represents the magnetic vector potential (see Chapter 3) we search for values having prescribed in advance the boundary conditions: the potential or the normal derivative of it. The first value is introduced directly while the second one again needs to be approximated in the sense of formula (4.3). This means, in other words, that the order of accuracy decreases since u at internal nodes has an approximation of $O(h^2)$ order while at the boundary the approximation is of $O(h)$ order. This is a drawback of the method but in spite of that the finite difference method is still attractive and useful. Of course, one can modify the difference schemes and this is done in contemporary works (Wiak 1985, 1987). Also one can adopt the finite differences to, say, variational formulation, but then one should be careful since the derivatives occurring in the formulation are of the first order, which decreases the accuracy of calculation. All remarks written above are also valid for the three-dimensional case but the boundary conditions there are different as has been shown in Chapter 3.

There is always the problem of how to solve the system of equations but this problem is common to all methods and lies beyond the scope of this book. The reader should consult one of the many books which cover the problems of linear algebra and numerical solutions of matrix equations.

Before we leave the quasi-stationary case let us point out that the finite difference method has one more drawback: in its most simple variant the approximation parameter h which creates the grid should be uniform (h equals constant). As a result the eddy current density and magnetic field are not distributed uniformly. Thus, one can choose the parameter h to be very small in order to cover all places of interest but this increases the dimension of the problem to a great extent. Alternatively, one can assume a varying parameter h which, in turn, causes some difficulties in programming. Moreover, if the latter way is adopted the problem still remains as we do not know *a priori* how to vary this parameter from place to place.

The situation is even more difficult when we deal with the transient state, that is, time-dependent problems. Then the operator L has the following form

$$L = \nabla^2 - \alpha^2 \frac{\partial}{\partial t} \tag{4.8}$$

which introduces the problem of the approximation of the time derivative. The problem has been investigated for decades since Crank and Nicholson in 1947 as well as Paceman and Rachford in 1955 gave the foundations of the finite difference algorithms (see Potter 1973). Simply speaking one should correlate both time and space steps in order to ensure the convergence of the method. Let us say that the time considered is divided into n intervals (t_{i-1}, t_i). Then we can substitute the time derivative by some difference scheme, that is, by the division of increments of function at the interval Δt. The question arises then as to what is the basic value of the function for the remaining part of the equation. To answer this question we introduce here, in the same way as was done in Chapter 3, the so called θ-method which gives the form of the difference scheme:

$$u = (1-\theta)u^{n-1} + \theta u^n \tag{4.9}$$

One can easily notice that choosing $\theta = 0$ or 1 we reach the first or the second limit of the time interval, respectively. Choosing in turn the value of θ between 0 and 1 we take, as the difference scheme of space, the value of function inside the interval. This choice is of consequence because it can lead to the situation in which the process can be either convergent or non- convergent. What is more, the choice of θ decides also the order of error.

For the one-dimensional case one can write the following difference equation:

$$(1-\theta)L_h u_{j-1} + \theta L_h u_j = \alpha^2 \frac{u_j - u_{j-1}}{\Delta t} \tag{4.10}$$

where L_h is as in eqn (4.6).

As was mentioned before the convergence and the stability of this equation depends on the mutual relation between h , Δt, and θ. Detailed analysis of the above equation shows that for $\theta \in [0.5,1]$ it is stable and converging unconditionally while for $\theta \in [0,0.5)$ the additional condition should be fulfilled, namely

$$r = \frac{\Delta t}{\alpha^2 h^2} \leq \frac{1}{2(1-2\theta)} \tag{4.11}$$

Although the explicit scheme (θ=0) seems to be very convenient as it does not require to solve the system of linear equations, condition (4.11) strongly limits its usefulness. Choosing θ belonging to $[0.5, 1)$ we focus our attention on the truncation error. It can be proved that the choice θ=0.5 gives the order of the truncation error $O(h^2 + \Delta t^2)$ while for any other value this error is $O(h^2 + \Delta t)$. It is, therefore, beyond discussion that this choice is the best and it is known in applied mathematics as the Crank–Nicholson scheme. Finally substituting θ=0.5 to eqn (4.9) and showing operator L_h in the *explicit* form one obtains:

$$\begin{aligned} u_{i-1,j+1} + (-2 - 2/r)u_{i,j+1} + u_{i+1,j+1} = \\ -u_{i-1,j} + (2 - 2/r)u_{i,j} - u_{i+1,j} \end{aligned} \tag{4.12}$$

for $i = 1, 2, 3 \ldots N(Nh = l)$ and $j = 1, 2, 3 \ldots M(M\Delta t = T)$.

Of course, for the initial values of i and j we have the boundary conditions and the initial condition, respectively. It is indeed a serious problem in the finite difference method how to introduce the boundary conditions of Neumann's type. Using formula (4.3) one looses the global accuracy there since these finite differences are of order one. On the other hand, the boundary conditions of Dirichlet's type are introduced automatically as the values of function u on the boundary can be moved to the right-hand side of eqn (4.12). Introducing the initial condition is immediate as the computational process is of a recurrence nature so that it uses the previous time step to solve for the present while the first step is just the time $t = 0$ at which the value of the function u is prescribed. Equation (4.12) can be written in the matrix form and is as follows:

$$(\boldsymbol{A} - 2/r\boldsymbol{I})\boldsymbol{u}^{n+1} = (-\boldsymbol{A} - 2/r\boldsymbol{I})\boldsymbol{u}^n - \boldsymbol{B}^{n+0.5} \tag{4.13}$$

where $\boldsymbol{A}$ is the three-diagonal matrix, $\boldsymbol{u}$- the solution vector, $\boldsymbol{I}$-the unit diagonal matrix and $\boldsymbol{B}^{n+0.5}$ the boundary and initial conditions vector.

Passing now to the two-dimensional problems one can see that the main problems still remain, that is, the problems of stability and convergence as well as truncation error. We can adopt here the Crank–Nicholson scheme,

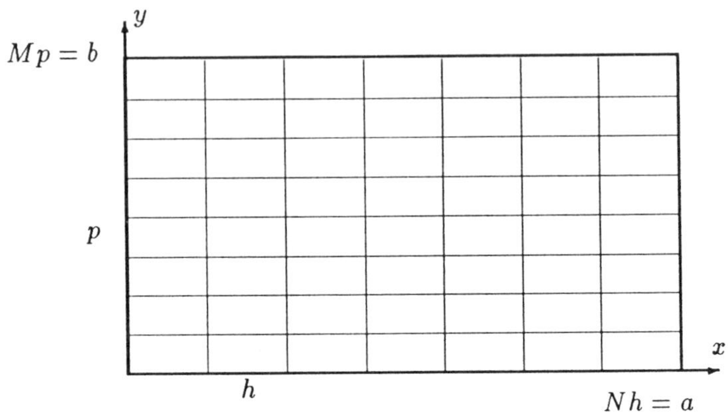

FIG. 4.1. The finite difference mesh

which will ensure the conditions requested. But it is well-known from the mathematical books (for example, Smith 1967) that for the case considered a better approach can be adopted. It is the so-called ADI method (*alternating directions implicit*), which guarantees that the calculation will be optimal.

Recently new mutations of this approach have appeared but we shall show here only the classical variant. Assume the rectangular domain $(0, a) \times (0, b)$ and introduce the uniform division towards each side: $Nh = a$ and $Mp = b$ (Fig. 4.1). Keeping also the uniform time step we can write the following two difference equations for the successive time steps:

for the time interval $(k\Delta t, (k+1)\Delta t)$

$$\begin{aligned}\frac{1}{h^2}(u_{i+1,j,k+1} &- 2u_{i,j,k+1} + u_{i-1,j,k+1}) \\ &+ \frac{1}{p^2}(u_{i,j+1,k} - 2u_{i,j,k} + u_{i,j-1,k}) \\ &= \frac{\alpha^2}{\Delta t}(u_{i.j,k+1} - u_{i,j,k}) \end{aligned} \tag{4.14}$$

for the time interval $((k+1)\Delta t, (k+2)\Delta t)$

$$\begin{aligned}\frac{1}{h^2}(u_{i+1,j,k+1} &- 2u_{i,j,k+1} + u_{i-1,j,k+1}) \\ &+ \frac{1}{p^2}(u_{i,j+1,k+2} - 2u_{i,j,k+2} + u_{i,j-1,k+2})\end{aligned}$$

$$= \frac{\alpha^2}{\Delta t}(u_{i,j,k+2} - u_{i,j,k+1}) \tag{4.15}$$

The first equation is of *implicit* nature as to the x direction and is of *explicit* nature as to the y direction, while the second equation behaves inversely. The method is stable and convergent unconditionally. Thus, at each time step we have to solve systems of linear equations having the dimensions $M - 1$ or $N - 1$ with matrices having a three-diagonal band structure. The boundary conditions are moved, as it was described for the one-dimensional case, to the right-hand side. Here also the problem of a lower accuracy occurs, namely the whole procedure has the truncation error $O(h^2 + p^2 + \Delta t^2)$ while on the boundary, if we use the non-symmetric finite difference to approximate the Neumann condition, the accuracy is of order one less. To ensure global accuracy it is necessary to use the symmetric finite difference, taking the point beyond our domain in question, but this increases the number of equations. It seems to be more profitable to accept the local loss of accuracy in return to the smaller number of equations. To make the ADI method more clear let us write the algebraic equations (4.14) and (4.15) in a matrix form:

$$(\boldsymbol{A}_1 + \boldsymbol{I})\boldsymbol{u}_j^{k+1} = \boldsymbol{T}_j^k - \boldsymbol{B}_j^{n+1} \tag{4.16}$$

where $j = 1 \ldots N - 1$

$$(\boldsymbol{A}_2 + \boldsymbol{I})\boldsymbol{u}_i^{k+2} = \boldsymbol{T}_i^{k+1} - \boldsymbol{B}_i^{k+2} \tag{4.17}$$

where $i = 1 \ldots M - 1$.

Matrices $\boldsymbol{A}_1$ and $\boldsymbol{A}_2$ are three-diagonal and each row and column (excluding, of course, the first and the last ones) is as follows:

$$\{r_{1,2}, -2r_{1,2}, r_{1,2}\} \tag{4.18}$$

where $r_1 = \Delta t/(h^2\alpha^2), r_2 = \Delta t/(p^2\alpha^2)$.

Vectors $\boldsymbol{T}$ have the form of

$$\boldsymbol{T}_j^k = \{-r_2 u_{i,j+1,k} + (2r_2 - 1)u_{i,j,k} - r_2 u_{i,j-1,k}\} \tag{4.19}$$

$$\boldsymbol{T}_i^{k+1} = \{-r_1 u_{i+1,j,k+1} + (2r_1 - 1)u_{i,j,k+1} - r_1 u_{i-1,j,k+1}\} \tag{4.20}$$

Vectors $\boldsymbol{B}$ contain the boundary conditions at the time $k+1$ and $k+2$, respectively, while matrix $\boldsymbol{I}$ is the unit (diagonal) matrix.

The above procedure has a recurrence form so that the final result depends on the initial condition imposed.

The procedures are somehow optimal for the one- or two-dimensional problems. For the three-dimensional transient problems the finite difference method is no longer used and hardly occurs in the literature. Even

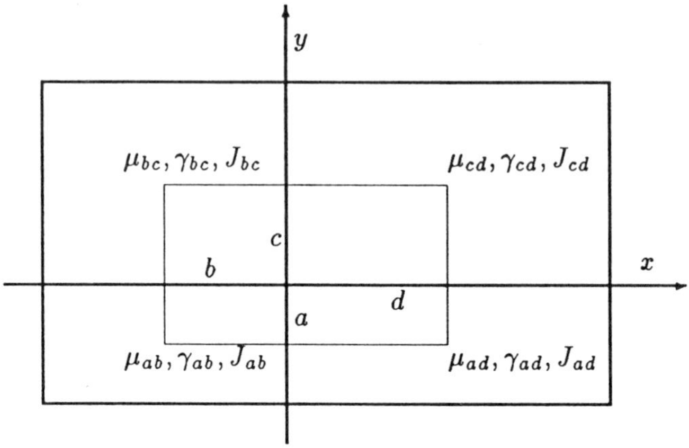

FIG. 4.2. Element of mesh consisting of four cells

for quasi-stationary problems the finite difference method is used very seldom in the three-dimensional analysis. Nevertheless it is sometimes very convenient to use this method as it is very clear for simple structures and does not require sophisticated mathematical tools.

The finite difference method can also be adopted in integral or variational formulations. The first formulation especially is worth noting since it allows us to overcome the problem mentioned above, namely that of non- uniform mesh.

Let us describe it according to Stoll (1974), who made use of Stokes' theorem.

Consider the eddy current area analysed by a five-point scheme (Fig. 4.2). The parameters in each cell can be different, which is also an advantage of the approach. Making use of the magnetic vector potential one can write eqn (2.1) in the form

$$\mathrm{curl}_z\left(1_x\nu\frac{\partial A}{\partial y}-1_y\nu\frac{\partial A}{\partial x}\right)=J_e-i\omega\gamma A \tag{4.21}$$

where $\nu=\mu^{-1}$.

Now, integrating both sides of the above equation over the domain bounded by a,b,c,d and making use of Stokes' theorem to the left-hand side, one obtains

$$\int_{abcd}\left(1_x\nu\frac{\partial A}{\partial y}-1_y\nu\frac{\partial A}{\partial x}\right)\cdot 1_l dl=\int_a^b\int_c^d (J_e-i\omega\gamma A)\,dS \tag{4.22}$$

The analysis of all particular contributions to the line integrals leads to the formula expressing the value of A at the central point 0 by the values at surrounding nodes:

$$A_0 = \frac{\alpha_1 A_1 + \alpha_2 A_2 + \alpha_3 A_3 + \alpha_4 A_4 + I_0}{\alpha_1 + \alpha_2 + \alpha_3 + \alpha_4 + j\omega Q_0} \tag{4.23}$$

where

$$\begin{aligned} \alpha_1 &= (a\nu_{ac} + b\nu_{bc})/2c \\ \alpha_2 &= (c\nu_{ac} + d\nu_{ad})/2a \\ \alpha_3 &= (a\nu_{ad} + b\nu_{bd})/2d \\ \alpha_4 &= (c\nu_{bc} + d\nu_{bd})/2b \\ I_0 &= acJ_{ac} + adJ_{ad} + bcJ_{bc} + abJ_{ab} \\ Q_0 &= ac\gamma_{ac} + ad\gamma_{ad} + bc\gamma_{bc} + ab\gamma_{ab} \end{aligned}$$

The method is also valid for other field quantities satisfying Helmholtz's equation, for example, if the magnetic field strength is used instead of the magnetic vector potential. It is obvious that for uniform mesh the above formula is the same as that derived from Taylor's expansion (see eqn (4.7)).

Applications of the FDM

The finite difference method considered here is not as popular as others methods, its popularity having declined in the last few decades. Let us notice that in the early era of computers, that is, in the 1950s and in the early 1960s, it was the only method existing. At that time the FDM was of great use in electrical engineering (Erdelyi and Fuchs 1965). As was mentioned the FDM has been used in problems with rather simple geometry, so that the mathematical models of technical devices were simplified to a great extent. It was, among others, Stoll (1974) who used the method in the eddy current problems but his examples are also geometrically simple and show the features of the method rather than calculation of real devices. Some researchers extended the life of the FDM throughout the 1980s and the reader is referred to Müller's paper (Müller 1983) as well as to the works of Wiak (Wiak 1985, Wiak and Zakrzewski 1987). In the latter one can find an interesting solution of electromechanical converter problems. Also, there is an interesting combination of the FDM and circuit analysis which allows determining transient circuit parameters of the problem considered like resistances or reactances.

4.2.2 *Method of finite sums (collocation method)*

Like the finite difference method which represents the *operator approximated* methods for differential formulation of eddy current problems, the method of finite sums represents the same idea but for integral operators. Indeed, coming back to Section 3.2.2 the integral approach is given by the Fredholm's integral equation of the second kind (see eqn (3.53)). Although the integral formulation is valid for three-dimensional cases we shall limit ourselves to two-dimensional case: the method presented here seems to be inadequate for the solution of three-dimensional cases, but is well-suited to two-dimensional cases. The procedure for the numerical solution of it is given below.

Before we start describing it let us notice that like the finite difference method, the method in question has some physical interpretation, namely it pictures the network of mutual inductances linking sub-domains of the conductor considered.

The integral Fredholm equation in terms of operators has the form:

$$g + \alpha^2 Lj + j = 0 \tag{4.24}$$

where operator L is

$$L \equiv \int_\Omega K(x, y, x', y')d\Omega \tag{4.25}$$

and $\alpha^2 = i\omega\mu\gamma$.

Function K is expressed in logarithmic form, if free space is considered, or has more complex form for internal boundary value problems. Function g represents, in turn, the grad ϕ (electric field) connected with eddy current flow which was discussed in Section 3.2. Some authors (Szymanski 1982) treat this function as resulting from Faraday's law, that is, this is the value of current density at the point of reference which is unknown.

Regardless of physical meaning, assigned to it in the two-dimensional case, this function is constant and is an additional unknown determined from eqn (3.54), which expresses continuity of eddy currents.

Take the conductor of cross-section surface Ω carrying current I. The cross-section of the conductor is divided into N sub-domains Ω_i (Fig. 4.1). Then the operator L can be approximated by the sum as follows:

$$\int_\Omega f(x)dx \mapsto \sum_{n=1}^{N} \int_{\Omega_n} f(x)dx \tag{4.26}$$

Now, assumimg that the integrand is constant within each elementary sub-domain Ω_i one obtains the following discretized equation:

$$g(x_i, y_i) + \sum_{n=1}^{N} j_n K_n(x_i, y_i, x_n, y_n) + j(x_i, y_i) = 0 \tag{4.27}$$

where

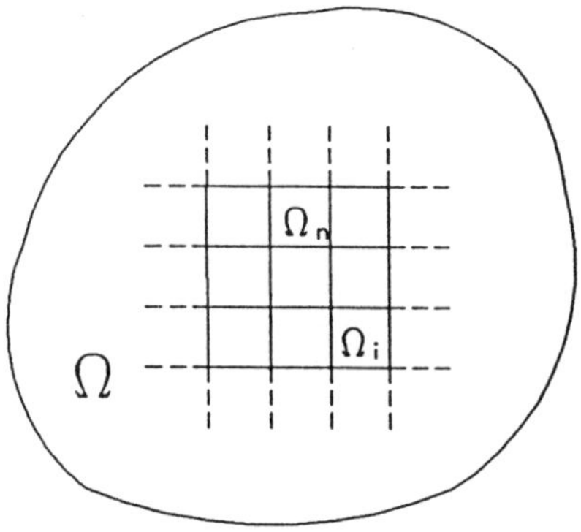

FIG. 4.3. Sub-division of conductor

$$K_n = \int_{\Omega_n} K(x_i, y_i, x_n, y_n) \tag{4.28}$$

Since K is determined *a priori* the problem is how to calculate the above integrals. The simplest way is to take the values of integrands as constant but this may lead to great error. If the sub-domains are rectangular the integrals for many cases (at least those of logarithmic type) can be calculated analytically. For arbitrary shapes the numerical integration should be adopted and then one must be aware that singularity may occur when $i = n$. The methods of dealing with singular integrals are common to all integral approaches and are mainly solved with analytical expressions. This problem will be discussed in more detail later on when the boundary element method is described. Physically the process of determination of singular integrals signifies searching for self-inductances. Having calculated them, a system of algebraic equations is established:

$$\{g\} + [K'] \{j\} = \{0\} \tag{4.29}$$

where

$$K'_{k,l} = \delta_{k,l} + \alpha^2 K(x_k, y_k, x_l, y_l) \tag{4.30}$$

$\delta_{k,l}$ is Kronecker's delta:

$$\delta_{k,l} = \begin{cases} 1 & k = l \\ 0 & k \neq l \end{cases}$$

As was mentioned above the unknown g is an additional unknown, thus an additional equation is needed and its discretized form is as follows:

$$\{N\}^T \{j\} = I \tag{4.31}$$

where $N_i = \Omega_i$ and I is the total current which must be known *a priori*. Finally the matrix equation has the form:

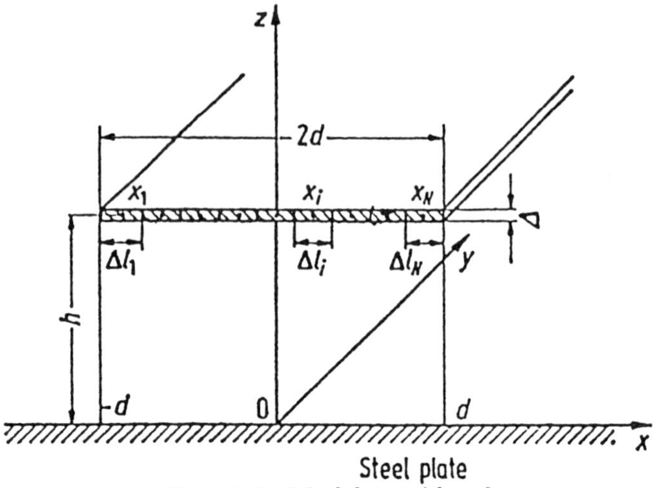

FIG. 4.4. Model considered

$$\begin{bmatrix} \boldsymbol{K}' & \{1\} \\ \{N\}^T & 0 \end{bmatrix} \begin{Bmatrix} j \\ g \end{Bmatrix} = \begin{Bmatrix} 0 \\ I \end{Bmatrix} \tag{4.32}$$

Let us stress that the calculation of the coefficients of matrix K' may be difficult. However, if the sub-domains are rectangular the integration results in analytical formulae.

Although the method is presented for one conductor only, its use can be extended to a multi-conductor system; the number of equations is M times greater then, where M is the number of conductors. Also, the method can be employed in more complex cases, for example, the problem of a long-enough current-carrying strip placed above the conducting half-space. Then the magnetic vector potential has two components: the first one is of logarithmic type and the second one results from eddy currents flowing in half-space, thus one has to change the kernel of the integral equation in order to consider this fact. The example of such a calculation has been presented by Szymanski (1982). The model he considered and the results obtained are presented in Figs. 4.4 and 4.5.

The method in question can also be applied to transient problems. However, since the difference scheme needs to be adopted, the method is not as effective as it is in the quasi-stationary case. In a quite similar way one can solve the problem of axisymmetric nature. Then, of course, the integrands are much more complicated and usually expressed in terms of elliptic integrals.

The method presented is similar if not the same as the collocation method but the latter is mostly derived from the weighted residuals method or from the method of moments (Harrington 1978). Indeed, anticipating the next section one can take for the solution of eqn (4.24) the following

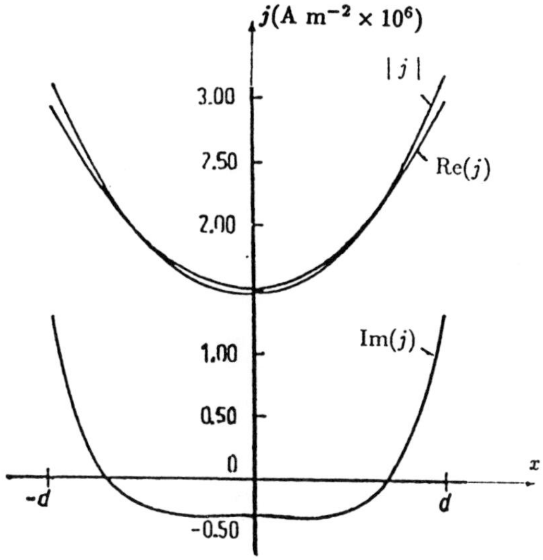

FIG. 4.5. Eddy current distribution

approximation of function j:

$$j(x,y) \simeq \sum_{n=1}^{N} j_n \phi_n(x,y) \tag{4.33}$$

The test functions ϕ_n are of the form:

$$\phi_n(x,y) = \begin{cases} 1 \text{ if } x,y \in \Omega_n \\ 0 \text{ otherwise} \end{cases} \tag{4.34}$$

Substituting the above formulae in the operator equation (4.24) and using the continuity of current, one obtains just the same matrix equation. In our classification, this belongs to the group of methods with approximation of function. It seems, however, that the approximation of function is so simple that it is better to give it here.

4.3 Approximation of function

4.3.1 *Method of weighted residuals*

This method is the groundwork for all the methods based on approximation of function. It would be difficult to say who was the founder of this approach as it is deeply rooted in the mathematics, but undoubtedly it was Finlayson (1972) who gave the basis of it, especially for technical applications. The idea of the method is as follows:

Let us take the operator equation

$$Lu = f \tag{4.35}$$

Approximating the function u in the n-dimensional space we obtain

$$u_h = \sum_{i=1}^{n} \beta_i \varphi_i \tag{4.36}$$

where β_i are the unknown coefficients and φ_i are linearly independent functions.[1] Functions φ_i are called basis functions. Substituting now the approximation of u to the operator equation (4.35) and comparing the two results for operator L, we obtain the error of approximation:

$$r = Lu - Lu_h \tag{4.37}$$

Now we can minimize the error in the average sense, namely we introduce the weighted functions (or test functions) v_i and calculate the integral weighted average over the domain of the operator L operation. This yields:

$$\int_\Omega r v_i d\Omega = 0 \tag{4.38}$$

Taking again Helmholtz's operator

$$L \equiv \nabla^2 - \alpha^2 \tag{4.39}$$

and introducing both (4.36) and (4.38) one finally obtains:

$$\int_\Omega \left(\nabla^2 u - \alpha^2 u - \nabla^2 \sum_{i=1}^{n} \beta_i \varphi_i + \alpha^2 \sum_{i=1}^{n} \beta_i \varphi_i \right) v_i d\Omega = 0 \tag{4.40}$$

The choice of function v_i determines the method of calculation and establishes the numerical method. Finally, taking into account (4.35), we have

$$\int_\Omega \nabla^2 \left(\sum_{i=1}^{n} \beta_i \varphi_i \right) + \alpha^2 \left(\sum_{i=1}^{n} \beta_i \varphi_i \right) v_i d\Omega = \int_\Omega f v_i d\Omega \tag{4.41}$$

Looking at eqn (4.41) from a conceptual point of view, it is easy to notice that it is similar to the variational form presented in Chapter 3. Indeed it is almost the same because the backgrounds of both are the same. The difference consists in the fact that the variational form was derived under the assumption that the operator L is self-adjoint and positively

[1] This means that $\sum \alpha_i \varphi_i = 0$ implies $\alpha_i = 0, \forall i$.

determined. In this case the operator is of an arbitrary nature and then the correspondence between this methods and the 'weak' formulation (see Section 3.2.3) exists.

The procedure presented above leads to the algebraic form which is as follows:

$$[\boldsymbol{A}]\{\beta\} = \{\boldsymbol{F}\} \tag{4.42}$$

The coefficients of matrix $\boldsymbol{A}$ are expressed by:

$$a_{ij} = \int_{\Omega} L(\varphi_j) v_i d\Omega \tag{4.43}$$

where $\{\beta\}$ is the vector of unknown coefficients and $\{\boldsymbol{F}\}$ is the vector of the right-hand side which represents the exciting sources. The main problem of the method is the calculation of the matrix coefficients, which is not as simple as in the finite difference method. But it may become simpler if the approximation expansion of function u as well as the weighted function v_i are arbitrary. As was mentioned above, we can choose both of them freely by keeping only the necessary mathematical features as to integrability and differentiability of functions. Taking functions v in the form of Dirac or quasi-Dirac functions one obtains the so-called collocation or point-matching method. Taking the weighted function the same as the basis function ($\varphi_i = v_i$) in the approximating series, one obtains the Galerkin[2] method which definitely corresponds to Rayleigh–Ritz[3] method if the operator L is positively determined and self-adjoint. In this method there also exist various possibilities and a compromise consists in the choice between the calculation of matrix coefficients and the quality of approximation, which, in turn, is strongly linked to the number of equations.

There are two extreme states:

- The first one consists in the choice of the basis and weighted functions to be bi-orthogonal. Then the matrix will be diagonal and the final calculation becomes very easy. Unfortunately it requires the coefficients of the matrix to be calculated as integrals over the whole domain, which constitutes a problem in itself.
- The second extreme state is that we cover the whole domain by very small elements and the basis function is kept as simple as possible within each element. Then, keeping the approximation satisfactory,

[2]Boris G. Galerkin (1871-1945): Russian engineer working on the theory of elasticity, famous for the approximation methods which were invented in 1915.

[3]John Wiliam Rayleigh (1842-1919): English phycisist working on the theory of oscillations; actually he invented the method towards the end of the nineteenth century. Walter Ritz (1878-1909): German mathematician; he described the method in 1908 but his name appears on the method with that of Rayleigh probably because of his dramatic death (he died of tuberculosis at the age of 31).

the number of equations will increase while the calculation of matrix coefficients becomes very easy. This may lead to sub-region collocation method.

Between these two extremes one can localize the mid-state in which one uses some kind of a quasi-orthogonal basis function and divide the whole region into small elements. This approach is recognized as the finite element method and will be the subject of the next section.

4.3.2 *Finite element method*

There have been many papers and books devoted to the finite element method during the last two decades. Most of them, however, concern the Laplace and Poisson equations, which from our point of view are less interesting. We therefore focus our attention immediately on the Helmholtz and parabolic equations. Nevertheless, some introductory remarks as to the nature of finite elements need to be given.

As was mentioned, the method is a variant of the Galerkin method and still more general, of the weighted residuals method. However, the method can be derived from the variational approach and then the Rayleigh–Ritz method should be applied. At the earlier age of the finite element method this approach was mostly used. Nevertheless, Galerkin's method is more general and for some problems is equivalent to the Rayleigh–Ritz method. For these problems the final algebraic forms are the same when using the two approaches.

The method is uniquely determined by the basis (weighted) functions which have two basic features:

- the coefficients of the approximating series expansion are of the interpolation type
- the basis functions have supports[4] considerably smaller than the domain considered.

The second feature introduces the topology of the mesh while the first one determines in some way the form of these functions.

In the most popular variant of the method one meets the triangle division and the linear basis functions spanned over these elements. In practice we have the function which represents a pyramid based on a polygon being the sum of triangles (elements) and a common vertex at the node considered. The application of such an approach makes the matrix with the coefficients relatively easy to compute while the matrix is of band structure or, at least, sparse.

To obtain a well-conditioned matrix the elements should be commensurable, that is, the ratio of their corresponding sides must be of the same order for all elements. This requirement, however, is somehow difficult to

[4]support of $f(x), x \in \Omega$, is equal to $\Omega_s \subset \Omega$, if $f(x) \neq 0$ for $x \in \Omega_s$ only.

fulfil in eddy current analysis. The reason is that the field quantity distribution is or can be very non-uniform. It is the skin effect which causes such a situation. Then, in order to examine the eddy currents flowing just in the skin layer, one has to use very small elements which are not commensurable with the ones inside the region considered. A similar problem appears when one deals with any other sub-domain which is exceedingly 'flat' (for example, the air gap in a reactor or electrical machine). As long as the problem is of geometrical nature only there are some approaches which can be used to overcome this difficulty. But in eddy current analysis the problem is physical since the skin layer is established by the phenomenon itself and not by geometry only. However, in spite of this and also other difficulties that will be discussed later, the finite element method is still widely used. That is why we shall show the mathematics of the method as it is being applied to the eddy current problem.

At first, consider the quasi-stationary case, that is, the sinusoidal eddy currents. Then the Helmholtz's equation holds and the integro-variational form follows from the weak formulation or from the variational procedure (in this case both approaches are interchangeable) as well as from the method of weighted residuals (after introducing Green's formula) and is as follows:

$$\int_{\Omega}\left(\sum_{i=1}^{d}\frac{\partial u}{\partial x_i}\frac{\partial v}{\partial x_i}-\alpha^2 uv\right)d\Omega=\int_{\Omega} fv\,d\Omega+\int_{\Gamma} v\frac{\partial u}{\partial n}d\Gamma \tag{4.44}$$

where d is the dimension of the problem and x_i is the i-th component of the space vector $\boldsymbol{x}$.

Equation (4.44) describes a physical situation, namely, introducing the magnetic vector potential in the place of function u, eqn (4.44) represents the relation of magnetic energy stored in the body and energy dissipated due to eddy currents.

Introducing the Galerkin procedure, that is, substituting u by an approximation series

$$u\cong\sum_{i=1}^{n}u_i\varphi_i(\boldsymbol{x}) \tag{4.45}$$

and choosing $v=\varphi_j(\boldsymbol{x})$, eqn (4.44) transforms itself to the following algebraic equation:

$$\sum_{j=1}^{n}u_i a(\varphi_i,\varphi_j)-\alpha^2(\varphi_i,\varphi_j)=(f,\varphi_i)+\sum_{k=1}^{n_b}u_k\left[\varphi_i,\frac{\partial\varphi_k}{\partial n}\right] \tag{4.46}$$

where φ_i are the basis functions described above and φ_k are the basis functions connected with the boundary nodes; $a(\varphi,\psi)$ is the bi-linear form defined as

$$a(\varphi, \psi) = \int_\Omega (\text{grad}\, \varphi \cdot \text{grad}\, \psi) d\Omega \tag{4.47}$$

while (φ, ψ) is the scalar product defined as

$$(\varphi, \psi) = \int_\Omega \varphi \psi d\Omega \tag{4.48}$$

and $[\varphi, \psi]$ is the scalar product again but determined on the boundary.

It is obvious that the matrix coefficients are functions of the coordinates of the point i and its surrounding space. As was mentioned above the matrix can be of a band structure if the geometry of the mesh is of rectangular shape. This is, however, a very rare case so in general there is no band structure but the matrix remains sparse.

The whole procedure in the finite element approach can be given by the four-step rule:

(1) the introduction of data and the discretization process (mesh generation);
(2) the creation of the matrix equation (integration process);
(3) solving the matrix equation;
(4) the output analysis (computation of the integral parameters as well as the derived quantities).

This process is commonly used for all problems tackled by the FEM, and all steps are described in several books (for example, Mitchell and Wait 1978).

Let us now focus our attention on the specific problems connected with eddy currents. As mentioned above, one of the main problems is the problem of discretization, that is, how to keep regularity of the division not knowing *a priori* the field distribution. This problem is much more serious in the transient state when the depth of skin layer varies with time. The solution of the problem may proceed by introducing the so called 'flat elements', which are one dimensional and have the same stored energy, as the real skin layer has, assuming a uniform field distribution in the layer (Nakata *et al.* 1981). Then the subdivision in the whole area remains commensurable and some nodes obtain an additional portion of energy. This procedure, however, requires the skin layer to be predetermined. For this reason this procedure is not useful in the transient state.

The second problem which is specific to eddy current analysis consists in the calculation of matrix coefficients. In the analysis of static problems, that is, when dealing with Laplace's or Poisson's equations, the matrix coefficients contain only integrals of function derivatives. It is obvious that these integrals are much simpler that those of the functions themselves. Looking at eqn (4.44) one can see that just the latter integrals occur in the eddy current problems. Even the linear approximation can make the

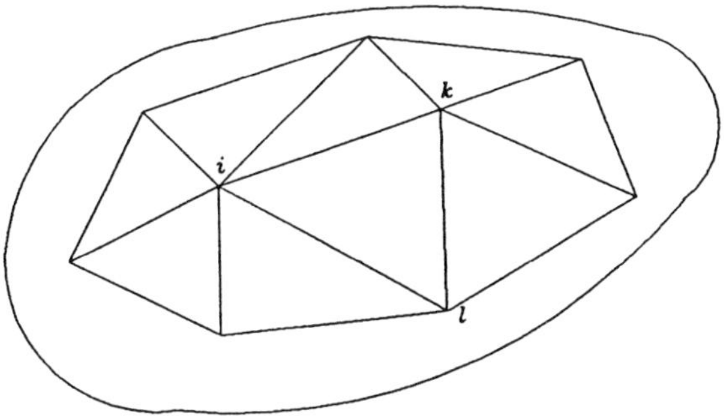

FIG. 4.6. Example of triangularization

calculation cumbersome. Of course, the approximation of higher order makes it even more difficult. It is a common source of difficulty for both quasi-stationary and transient problems. However, as has been mentioned repeatedly, in spite of all difficulties the finite element method is still very popular and has many followers. This results from the fact that nowadays very complex software for the FEM is available, which includes almost all two-dimensional problems as well as some of three-dimensional cases. This software includes such widely known packages as OPERA-2D, TOSCA, CARMEN, COMPEL, MAGNET, FLUX3D, EDDYCUFF, and the like. What is still not well-developed is the case of three-dimensional problems for both non-linear and transient cases.

In the following the method is described in detail and some pecularities of the method are given.

Let us take the domain Ω belonging to two-dimensional space. Assume that the space is regular enough so that the triangularization can be accomplished (Fig. 4.6). To each internal node i resulting from the discretization, the basis function φ_i may be assigned, which is equal to 1 at the node and zero at the surrounding nodes (Fig. 4.7).

In this particular case if the xy plane is considered, the basis function related to the node i and the triangle ikl is given as follows:

$$\varphi_i(x, y) = a_i x + b_i y + c_i \tag{4.49}$$

where the coefficients a_i, b_i, c_i are such that they guarantee the basic features of the basis function and are different for each particular triangle surrounding node i. Thus, they are as follows:

$$a_i = \frac{y_k - y_l}{2S}$$
$$b_i = \frac{x_l - x_k}{2S}$$

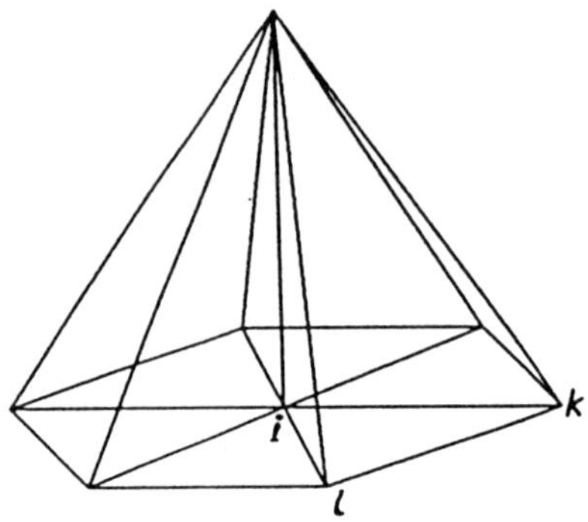

FIG. 4.7. The geometrical representation of the basis function

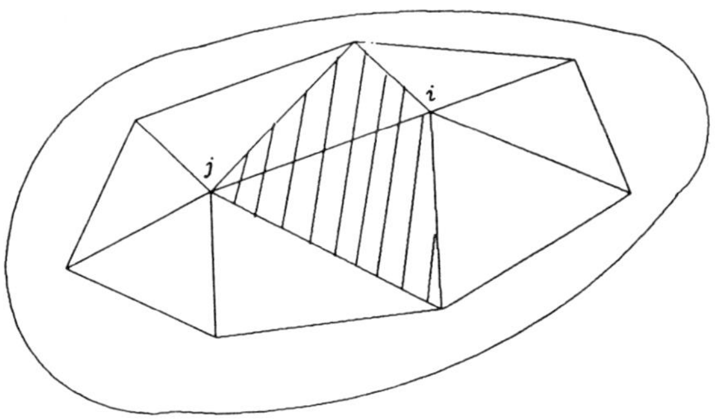

FIG. 4.8. Domain of integration

$$c_i = \frac{x_k y_l - x_l y_k}{2S}$$

where S is the area of the triangle ikl.

Such a choice of basis function means that the integral given in eqn (4.44) is limited to the area of a few triangles linking node i and some of its neighbouring nodes. Let us consider the relation between two neighbouring nodes i and j. In this case the only elements to be taken into account in the matrix coefficients indexed by i and j are those marked in Fig. 4.8.

Over these triangles one has to evaluate the following integrals:

$$I_1 = \int_{\Omega_{ij}} \left(\frac{\partial \varphi_i}{\partial x} \frac{\partial \varphi_j}{\partial x} + \frac{\partial \varphi_i}{\partial y} \frac{\partial \varphi_j}{\partial y} \right) dx dy \tag{4.50}$$

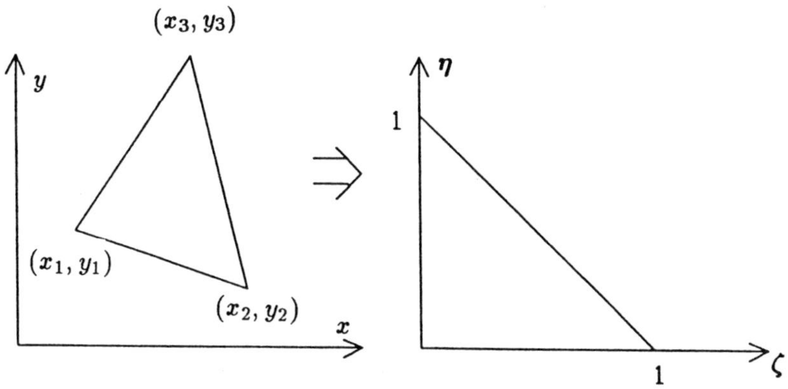

FIG. 4.9. Tranformation of triangle

$$I_2 = \int_{\Omega_{ij}} \varphi_i \varphi_j dx dy \tag{4.51}$$

where Ω_{ij} is the element linking i-th and j-th nodes.

The first integration is rather simple as the integrand is constant, while the second one requires computation. The integration can be done either analytically or numerically. In both cases the usual procedure consists in the linear transformation of any arbitrary triangle to the standard one (Fig. 4.9) which is expressed as follows:

$$x = A\zeta + B\eta + C \tag{4.52}$$

$$y = D\zeta + E\eta + F \tag{4.53}$$

Hence the above-written integrals are transformed into the following forms:

$$I_1 = \int_0^1 \int_0^{1-\zeta} f(A, \ldots, F, \zeta, \eta) \mid J(\zeta, \eta) \mid d\eta d\zeta \tag{4.54}$$

$$I_2 = \int_0^1 \int_0^{1-\zeta} g(A, \ldots, F, \zeta, \eta) \mid J(\zeta, \eta) \mid d\eta d\zeta \tag{4.55}$$

where $J(\zeta, \eta)$ is the Jacobian of the above transformation.

It should be repeated again that the above integrals can be solved analytically although the numerical approach can be successfully applied. This way one can calculate all the matrix coefficients, and finally the matrix equation becomes

$$\boldsymbol{Au} = \boldsymbol{B} \tag{4.56}$$

Note that the coefficients of matrix $\boldsymbol{A}$ are of complex nature which is not a problem for programming nowadays. In the past, however, attempts were made to remain in the domain of real numbers and then Helmholtz's equation was transformed to biharmonic by separating real and imaginary parts (Pawluk and Szweicer 1978).

The vector of right-hand side is made of two components. The first results from the excitation (force) function while the second from the boundary conditions. The computation of the first component is conceptually rather simple and requires the integration over triangles connected with each node:

$$B_i = \int_\Omega f\varphi_i dxdy \tag{4.57}$$

and the same procedure as above may be employed.

The second component needs some deeper considerations which are given below.

There are two approaches for introducing boundary conditions—both resulting from their type. Thus, the Neumann condition introduces itself in natural manner because it is included in integral form (see eqn 4.44). A far worse situation is encountered when the Dirichlet condition is given, which is the opposite case to the finite difference method. The introduction of the Dirichlet condition is made again in different ways. One of them, widely used, is that the whole region including boundary is considered and then matrix $\boldsymbol{A}$ contains the coefficients connected also with the boundary nodes. After that, the matrix is reassembled in some way, that is, the coefficients connected with the nodes where the value of the function is known are eliminated from the matrix or substituted by others which enforce the proper value of the function. In this way, however, difficulties arise connected with the conditioning of the resultant matrix. The other way consists in substituting the trial function v at the very initial step. Function v included in eqn (4.44) may be considered as a sum of two components: one vanishing on the boundary while the second is zero everywhere except on the boundary and its neighbourhood:

$$v = v_0 + v_b \tag{4.58}$$

where

$$v_0 \mid_\Gamma = 0$$

$$v_b = \begin{cases} 0 & \Omega \setminus \Gamma \\ f(x) & x \in \Gamma \end{cases}$$

In terms of the matrix expression this means that the whole matrix is confined to the internal nodes while the coefficients resulting from the second component of function v are moved to the right-hand side. Then the boundary nodes require also the computation of $a(u, v_b)$ and (u, v_b) and this concerns the nodes located in the first layer close to the boundary

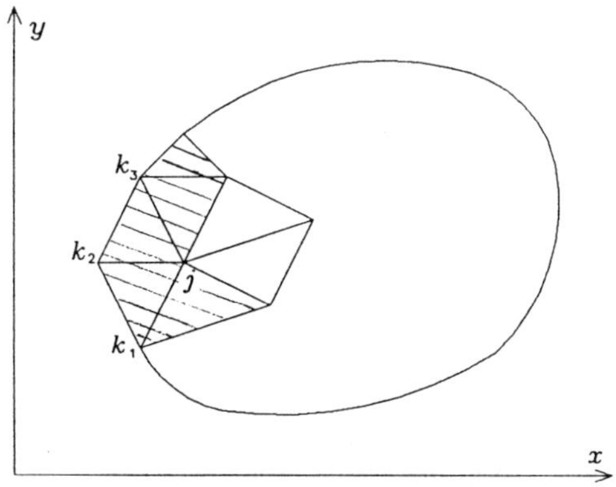

FIG. 4.10. Boundary layer of elements

(Fig. 4.10). In this way the matrix of the problem can be set up and the problem can now be solved. There are really no special differences in solving harmonic eddy current problems as compared with the well-documented static (Laplace's or Poisson's) problems.

Let us therefore concentrate our attention on the transient problems which are conceptually different in the finite element analysis. The transient problems bring new difficulties which are just typical for this kind of problems. The main difficulty is connected with the time analysis of those problems. Namely, if the problem is treated in terms of the magnetic vector potential, eddy currents result from the differentiation of the vector potential with respect to time. As the latter can have some non-physical oscillations in its time course, eddy currents may carry some unaccountable error. To avoid this difficulty the proper choice of time step as well as the difference scheme parameter should be made and this task is rather cumbersome from both a theoretical and a practical point of view. An excellent analysis of it has been presented in the paper by Zlamal (1976) as well as in the paper of one of the authors (Krawczyk 1983). Since it is the key-point of the method, this analysis will be briefly quoted below.

The finite element treatment of the parabolic equation expressed in weak form (see Section 3.2.3, eqn (3.66)) leads to the system of ordinary differential equations:

$$\boldsymbol{A}\frac{d\boldsymbol{\alpha}}{dt} + \boldsymbol{B}\boldsymbol{\alpha} = \boldsymbol{F}(t) \tag{4.59}$$

which, in this specific case, is the so-called 'stiff system'. Such a system, which is the subject of many mathematical monographs (for example, Lambert 1972), requires a special approach as the simple integration systems, like the Runge–Kutta method, fail here. The reason for this is that the spectrum of time constants in the time series is very wide. More precisely:

the system of ordinary differential equations is said to be stiff if the eigenvalues λ_i of the matrix $\mathbf{A}^{-1}B$ satisfy the following conditions:

(1) $Re(\lambda_i) < 0$

(2) $max \mid Re(\lambda_i) \mid >> min \mid Re(\lambda_i) \mid$

The use of the θ-method, mentioned in Chapter 3, is the simplest way to solve such a system of equations. Nevertheless, the choice of the parameter θ is to be made very carefully as the solution is sensitive to it. The deeper analysis of differential matrix equation (4.59) in context with the θ-method leads to the conclusion that this choice is rather difficult and at least not simple since some of the requirements are in contradiction, for example, the accuracy of the solution needs $\theta = 0.5$ while for the stability of solution the optimal value of θ is 0.122. Zlamal proposed to take $\theta = 0.2 \div 0.4$. In fact the parameter θ should not be constant, but vary according to the time considered. Note that taking $\theta = 1/3$ gives the one-dimensional finite element method (Galerkin approach) in time. Actually, there is no unique prescription how to choose the parameter and, more generally, what method of solving eqn (4.59) is the best. Lambert (1972) ends his analysis of *stiff systems* with the following sentence: '...the real test is the efficiency with which we can handle the resultant implicitness. Below we give some examples to illustrate the above considerations.

The simple example concerning one-dimensional penetration of the magnetic field into the conducting strip (see Fig. 2.1) shows this undesired effect. One considers here the one-dimensional diffusion equation, the analytical solution of which is well-known (see eqn (3.2)). The exciting field has the form of unit step function, conductivity γ is equal to 56 MSm^{-1} and the time step for the FEM solution is 0.01 s. In Fig. 4.11 the courses of relative errors at two space points (y=0.4b and y=0.3b) are shown with respect to time and one can see that the error strongly depends on the parameter θ which is the basis of the linear one-step method (Krawczyk 1983). One can observe here some oscillations which decrease with time.

It is obvious that the θ-method is not the only existing method. In applied mathematics one can find more sophisticated methods, such as multi-step methods or some specific ones which are appropriate for ordinary differential equations (for example, Pade's approximation). It should be repeated once again that the entire analysis of the problem considered can be found in papers concerning heat conduction theory and also in mathematical books (for example, Wilson and Nickel 1966; Zlamal 1976; Lambert 1972). From the electrical engineering point of view, however, the more complicated methods are not used and the θ-method is usually quite sufficient.

In spite of the difficulties encountered, the FEM still attracts the attention of analysts and the method is commonly used. The main idea is based on the 'weak formulation' shown in Chapter 3. The space approximation

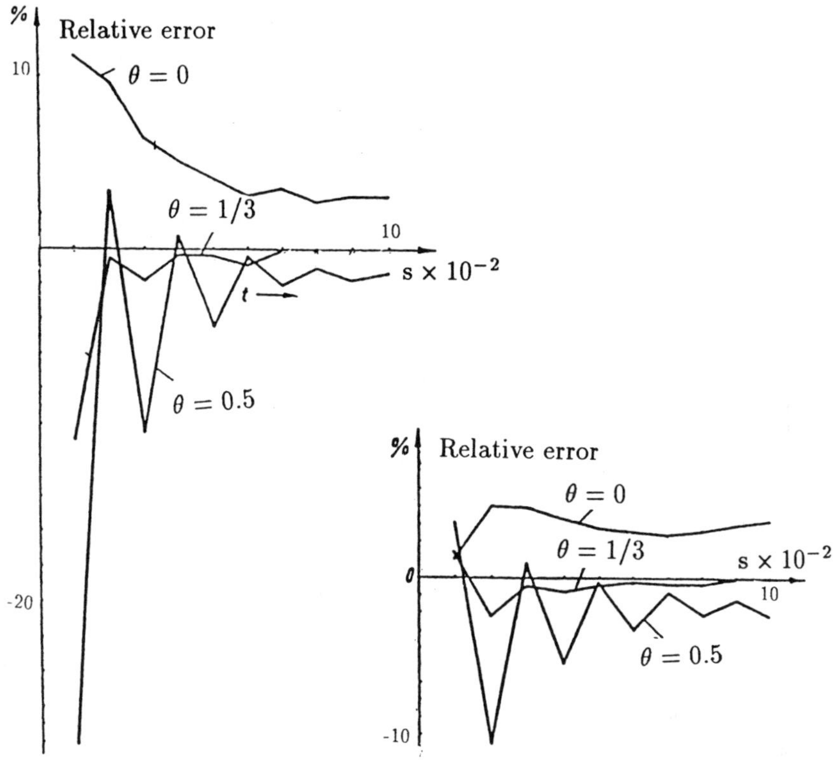

FIG. 4.11. The relative errors

is made in the same way as for the static cases and the time approximation recurrence process can be based on the simple linear one-step method or can be given in a more sophisticated way. The figure presented above (Fig. 4.11) shows the one-step linear approach which was mentioned already in Chapter 3.

The above presented methodology is connected with the two-dimensional analysis. The FEM may find even a larger field of applications in three-dimensional problems. Generally the philosophy of the method remains the same but the degree of complexity increases dramatically. Instead of flat elements one uses the geometrically spatial elements, that is, the three-dimensional ones, but as far as one considers the scalar fields, like thermal fields, the differences in methodology are not great. Usually the tetrahedral or brick elements are used. However, as was shown in Chapter 3 the eddy current three-dimensional fields are vectorial and, what is more they are coupled with scalar fields. Thus, the three-dimensional analysis may appear not so simple as the two-dimensional case. The details of geometrical aspects of the three-dimensional FEM are entirely described in the monograph by Ferrari and Silvester (1984).

It should be added here and also strongly stressed that the FEM has certain incompatibility with interface conditions resulting from the method of approximation. Namely, coming back to the two-dimensional problem it is easy to notice that when using magnetic vector potential and linear approximation on each element, the tangential component of the magnetic field on interfaces between elements is not continuous. Physically it can be supposed that an additional surface current occurs. It can be proved (Hammond 1989) that such a situation causes the total energy of the system to be lower than the exact one. This may have some important consequences. To avoid this situation one should use such an approach which would give an energy greater than the exact one. Indeed, applying the scalar potential one reaches that. This is accomplished at the price that all calculations are doubled. Another solution was developed recently and is shown below.

For the last two decades the finite element method has been used in electrical engineering in the form of the nodal variant. Since the late 1980s a new approach has been introduced by Bossavit (1988) and it is called *edge elements*. The approach results from the differential geometry and consists in introducing objects to the approximation space instead of nodes. Each spatial polyhedral element has three sets of describing features: nodes $\{i\}$, edges $\{e\}$ and facets $\{f\}$. Thus one can construct the approximation space based on one of these three sets. The conventional variant is based, of course, on the sets of nodes and this fact implies that the approximation space is of scalar nature (one-value space). Now, the use of edges as the 'nodal' points of approximation has as a consequence that the basis function is a vectorial one. To construct such a space one can use the so called Schauder basis known from the one-dimensional finite element.

Consider the tetrahedral element and take the nodes i and j linking the edge ij. Introducing the function λ_i to node i (Schauder's basis) defined as follows:

(1) linear

(2) $\lambda_i = \delta_{ij}$ (Kronecker's delta)

one can construct the edge basis function as follows:

$$\boldsymbol{w}_e = \lambda_i \nabla \lambda_j - \lambda_j \nabla \lambda_i \tag{4.60}$$

Then the approximation of an arbitrary function u can be established in the space spanned on the above functions $\boldsymbol{w}_e$, so that one can write:

$$u = \sum_{e=1}^{N_e} u_e \boldsymbol{w}_e \tag{4.61}$$

where N_e is the number of edges.

Note that the approximation is of vectorial nature. The edge element approach has fundamental and important consequences: *Maxwell's equations are exactly satisfied and boundary (interface) conditions fulfilled.*

Moreover, all physical requirements like the vanishing of the divergence of vector $\boldsymbol{B}$ and the continuity of tangential component of $\boldsymbol{H}$ are imposed here automatically by the edge elements themselves.

The edge elements are already adopted in real eddy current calculation and for an idea of how they work as well as some comparative analysis the reader is referred to the paper by Kameari (1989).

Applications of the FEM

The first technical use of the finite element method in electrical engineering occurred in the early 1970s. It is obvious that the primary applications concerned the static fields, that is, Laplace's or Poisson's equations. The first significant paper concerning FEM for eddy currents was that by Chari (1974). To complete the FEM story one should look at the book by Chari and Silvester (1981) and Ferrari and Silvester (1985). The eddy current applications grew up in the 1980s and the previous names should be supplemented by those from the Rutherford Appleton Laboratory (Trowbridge, Simkin, Emson) as well as from Japan (Nakata and his colleagues). The main groups of problems of interest to researchers are those concerning transformers and electrical machines. There is, of course, a need to compute eddy currents both in active and passive structural elements (windings, core, screens, various plates and bus-bars). Recently new devices began to be investigated. They are mainly devices which can be covered by the term *eddy current-based* devices. Here one can mention such devices as magnetic field concentrators, launchers of all types, magnetic forming tools, fast switchgear and the like. The usage of FEM in analysing these devices enables their performance to be developed.

A good example of the above mentioned applications is the electromagnetic pump designed by Bessho (1992).

4.4 Boundary element method

As was already mentioned in the discussion one formulations of electromagnetic problems, the boundary element method is always derived from the second Green identity. The boundary-integral formulation is discussed in Section 3.2.4.

Independently of whether quasi-stationary or transient cases are considered, the main idea of the BEM approach consists in choosing the approximation of function representing either the function itself or its normal derivative. There are some special requirements as to the compatibility between the order of the approximation of the function and the approximation of space, that is, in this case, the boundary.

Briefly speaking the order of approximation of the function should be at least the same as the order of space approximation. But in every case, for each space approximation one can take the constant approximation of function which is the simplest one. This remark is well-known from the theory of the FEM but taking a constant approximation of function seems to be specific for the BEM. Indeed, taking the linear elements in the two-dimensional case one can use either linear or constant approximation, and the latter is mostly used. Before we start considering the BEM technique let us notice that the constant approximation bears some inherent contradiction as it approximates with the same order both function and its normal derivative. For example, when one uses the magnetic vector potential A as a basis quantity one obtains in each element the constant tangential component of magnetic flux density and the zero normal component of magnetic flux density, which is not true. Also taking the higher order of approximation one obtains the linear distribution of tangential components and constant normal components which is not physically correct. To avoid this situation one can introduce different orders for each quantity but then the problem of matrix form appears as it becomes rectangular instead of quadratic. This causes, of course, considerable difficulties in a numerical operation. Thus, one should decide which way is preferable. It seems, however, that the physical contradiction is much less troublesome than the mathematical one. What is more, the difficulties of the physical contradiction decrease with the number of elements.

Thus let us consider the simplest case in two-dimensional quasi-stationary analysis when the boundary Γ of the domain Ω is divided into N straight-line elements $\{e_i\}$ and on each element the constant value of both boundary variables (function and its normal derivative) is assumed (Fig. 4.12).

The mathematical formulation is given in eqn (3.81) and introducing the approximation mentioned above one obtains the system of algebraic equations in the form:

$$0.5u_i + \sum_{j=1}^{N} u_j \int_{e_j} v'(r_{ij})d\Gamma = \sum_{j=1}^{N} \frac{\partial u}{\partial n} \int_{e_j} v(r_{ij})d\Gamma + \\ + \int_{\Omega} vf d\Omega \tag{4.62}$$

where $v(r)$ is the fundamental solution (*free-space Green's function*) of the Helmholtz equation and has the form

$$v(r) = \frac{1}{2\pi} K_0(\alpha r) \tag{4.63}$$

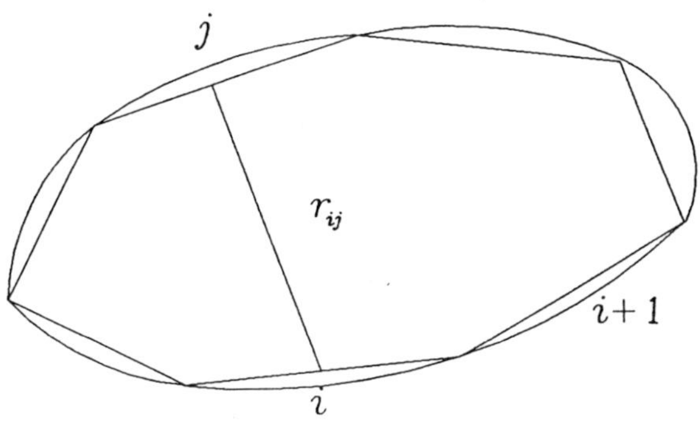

FIG. 4.12. The boundary element discretization

while its derivative is as follows:

$$v'(r) = \frac{\alpha}{2\pi} K_1(\alpha r) \frac{\partial r}{\partial n} \tag{4.64}$$

The problem is how to calculate the integrals especially when $i = j$. The latter will be discussed later on. For now let us consider the easier case $i \neq j$.

Using the analytical asymptotic formulae for the cylindrical functions one can work out the closed solutions of the integrals. But this seems to be more time consuming and the normal practice is to use for this purpose numerical integration. Usually, the Gaussian quadratures are used. Applying them one should transform an arbitrarily placed line segment into the one-coordinate line $(-1, 1)$ (Fig. 4.13)

Then the integrals occurring in (4.62) are as follows:

$$\int_{e_i} f(x, y) dl = \int_{-1}^{1} f(z) \mid J(x, y) \mid dz = \sum_{i=1}^{N_Q} A_i f(z_i) \mid J \mid \tag{4.65}$$

where A_i are the Gaussian quadrature weights, z_i denote the quadrature nodes and $\mid J \mid$ is the module of the transformation Jacobian. It is obvious that the order of quadrature influences the result of integration but a compromise should be found between the accuracy of calculation and the time required for it. Practice shows that the four-point quadrature is quite satisfactory.

Beside the normal difficulties connected with the numerical integration, a special problem appears which is related to the integration with singularity when $i = j$. The function K_0 has then the singularity of logarithmic

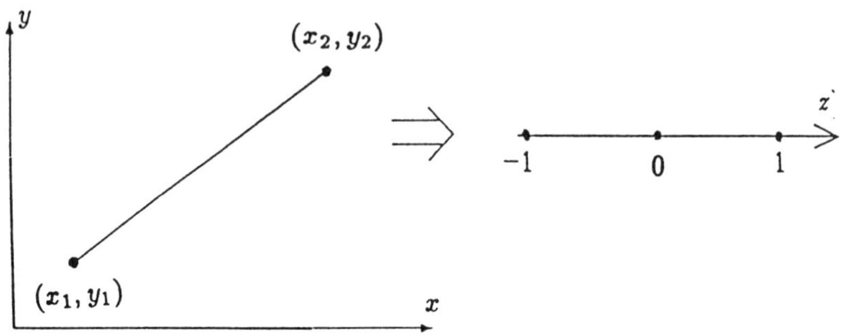

FIG. 4.13. The transformation of a straight segment $(x_1, y_1), (x_2, y_2)$

type in the area of integration so that the pure numerical integration fails. Let us notice here that there is no singularity problem for the function K_1 since when $i = j$ the integral is equal to zero. This is due to the orthogonality in r and n directions which causes that $\partial r/\partial n = 0$.

There are some special approaches to remedy this situation. Among them is the series expansion of the integrand, the approximation of the integrand by a well-integrable function or the analytical calculation of the integral, if this is possible. The latter is commonly used in the static problems where the integrand is expressed in terms of logarithms. In the case of the Helmholtz equation the series expansion can be employed. Limiting the series to a few terms (exactly three) the exact formula of integration can be obtained and it is as follows (Salon and Schneider 1981):

$$\int_{e_i} K_0(\alpha r_{ii}) d\Gamma = \frac{2}{\alpha} \left(\sum_{k=0}^{2} d_k \beta^k - \ln \beta \sum_{k=0}^{2} c_k \beta^{2k+1} \right) \tag{4.66}$$

where

$$\beta = \frac{\alpha e_i}{4}$$

and the coefficients of expansion are as follows:

k	c_k	d_k
0	2.0000	0.8456
1	0.6667	0.5041
2	0.1000	0.1123

Very close to the problem mentioned above is the problem of the so-called quasi-singularity which occurs in the BEM analysis. Namely, when analysing the internal points placed near the boundary one faces the small arguments of functions K_0 and K_1 and then the numerical integration may fail. To avoid this dangerous situation it is proposed here to use the classical technique consisting in adding and subtracting a part of the integrand which produces a quasi-singularity. In the two-dimensional case this part will be the logarithm and in the three-dimensional case it is the inverse radius r. The integrand minus a singular component can be integrated numerically without obstacles while the integrals of both functions introduced have analytical solutions.

In the early age of the BEM some authors (Schneider and Salon 1981) proposed to take as a boundary element integrals the product of a midpoint value of the integrand times the length of the element but such an approximation can cause a large error.

Having computed the integrals, the final matrix equation is now ready to be solved. The problem, however, may appear because the matrix diagonal may not be predominant. In this case all methods which require the diagonal predomination are excluded, including iteration methods. For this reason, the Gauss elimination method is preferred here.

Exemplary results for a two-dimensional problem with the use of the electric vector potential (stream function) has been reported, among others, by Schneider and Salon (1981). They compared their results with other methods and reached satisfactory agreement, although the matrix coefficients were calculated in an improper way (as mentioned above) except those of singular nature. This shows that the BEM is not especially affected by the integration process.

Quite new questions appear when one deals with the transient problems. Then, eqn (3.97) is valid and should be treated by the boundary elements. The space technique is rather similar if not the same as it was for the quasi-stationary case while the time procedure is entirely special. Equation (3.97) shows that substitution of the time integral $\int_0^t$ by the sum of integrals $\int_{t_{i-1}}^{t_i}$ enables the formulation of the functions which are the results of the integrals $\int_{t_{i-1}}^{t_i}$. These functions play a role of the fundamental solution and its normal derivative in the problem without time consideration. Thus, the problem can be treated now as quasi-stationary at each time interval (t_{i-1}, t_i). At this place let us quote the functions φ and β in order to make our further discussion more clear.

$$\varphi(\boldsymbol{x}; \boldsymbol{x}_i, t; t_i; t_{i-1}) = \int_{t_{i-1}}^{t_i} v dt = \frac{\alpha^4 d}{8\pi^2 r}[E_1(w_i) - E_1(w_{i-1})] \quad (4.67)$$

$$\beta(\boldsymbol{x}; \boldsymbol{x}_i, t; t_i; t_{i-1}) = \int_{t_{i-1}}^{t_i} \frac{\partial v}{\partial n} dt = [\exp(w_{i-1}) - \exp(w_i)] \quad (4.68)$$

where E_1 is the integral exponent expressed as follows:

$$E_1(x) = \int_0^x \frac{e^{-t}}{t} dt$$

$$w_n = \frac{\alpha^2 r^2}{4(t - t_n)}$$

$r = \mid \boldsymbol{x} - \boldsymbol{x}_i \mid$, d is the distance between the point $\boldsymbol{x}_i$ and the element of integration. However, all time intervals are linked and cannot be considered separately.

There are two ways of solving each interval in terms of time:

- The first one it is the 'step by step' method, which consists in using the results of a certain step as the initial condition for the next step. This requires the numerical domain integration which, in turn, means that a lot of internal points need to be calculated. The domain integral has an integrand which is a product of the initial condition and the fundamental solution v (see eqn (3.97)). Thus at each time step we have to solve the integral:

$$I = \int_\Omega u(\boldsymbol{x}, t_{k-1}) v(\boldsymbol{x}_i; \boldsymbol{x}, t; t_{k-1}) d\Omega \tag{4.69}$$

  where the function v is given by eqn (3.90).

  If the purpose of computation is such that a large amount of information about the internal region is desired then it is necessary to calculate the integral (4.69) and this is done numerically. On the other hand, if a user of the BEM is interested in a few points only then the necessity of numerical integration represents an amount of work disproportionate to the expected results.

- The second way of time treatment can be named the 'cumulative method' since it gathers all the information about boundary values during the whole time process, that is, from the initial instant ($t = 0$) to the time considered. This means that the calculations are conducted by taking previously calculated values as known and then the integrals connected to these values are moved into the right-hand side. Coming back to eqn (3.92) one can write for the time instant t:

$$\begin{aligned} .5\alpha^2 u(\boldsymbol{x}_i, t) &+ \int_\Gamma u(\boldsymbol{x}, t) \beta(\boldsymbol{x}, \boldsymbol{x}_i, t, t - \Delta t) d\Gamma \\ &- \int_\Gamma \frac{\partial u}{\partial n} \varphi(\boldsymbol{x}, \boldsymbol{x}_i, t, t - \Delta t) d\Gamma \\ &= \sum_{k=1}^{m-1} \int_\Gamma \frac{\partial u}{\partial n} \varphi(\boldsymbol{x}, \boldsymbol{x}_i, t, t_k, t_{k-1}) d\Gamma \end{aligned}$$

$$
\begin{aligned}
&- \sum_{k=1}^{m-1} \int_\Gamma u(\boldsymbol{x}, t_k)\beta(\boldsymbol{x}, \boldsymbol{x}_i, t, t_k, t_{k-1}) d\Gamma \\
&+ \sum_{k=1}^{m} \int_\Omega f\varphi(\boldsymbol{x}, \boldsymbol{x}_i, t, t_{k-1}, t_k) d\Omega \\
&+ \alpha^2 \int_\Omega v \mid_{\tau=0} u \mid_{t=0} d\Omega
\end{aligned} \tag{4.70}
$$

where φ and β are the kernels given by eqns (4.67) and (4.68), m is the number of time steps.

As one can easily notice such a procedure requires all the boundary values to be stored in the computer memory and be used every time at the next step. On the other hand, the integrals can be calculated at the beginning of the process and be also stored in the memory. When the time steps are non-uniform the array of these integrals should be three-dimensional while when time steps are uniform the array is two-dimensional.

The final choice between the two approaches is a very important matter as it gives the final shape of the computer program. It is not possible, on the basis of the above presentation, to say which of the two is the most preferable. This depends on the particular application.

Using again the straight-line elements and the constant approximation for the function u and its normal derivative q one can write the matrix form of boundary integral equation, which is as follows:

$$
\sum_{s=1}^{m} [H_{sm}]\{u_s\} = \sum_{s=1}^{m} [G_{sm}]\{q_s\} + \{F_m\} \tag{4.71}
$$

where the matrix coefficients are expressed by

$$
\begin{aligned}
h_{smij} &= \int_{e_j} \beta(r_{ij}, t_m, t_s, t_{s-1}) d\Gamma + \frac{1}{2}\alpha^2 u(\boldsymbol{x}_i, t_m)\delta_{ij}\delta_{sm} \\
g_{smij} &= \int_{e_j} \varphi(r_{ij}, t_m, t_s, t_{s-1}) d\Gamma \\
F_{mi} &= \int_0^{t_m} \int_\Omega f v(r_i, t_m, \tau) d\Omega dt + \alpha^2 \int_\Omega v(r_i, t_m, 0) u_0 d\Omega
\end{aligned}
$$

δ_{sm}, δ_{ij} are the Kronecker deltas.

Similarly to the Helmholtz equation, as discussed above, difficulties arise when calculating the singular integrals. The ways of tackling them are similar to those shown in the sinusoidal case. Here also the series expansion of the function E_1 is used. Then the analytical solution for integral g_{smij} is as follows:

$$g_{mmii} = \frac{\alpha^2 e_i}{4\pi}\left[2 - C - \ln w + \sum_{k=1}^{\infty}(-1)^{k-1}\frac{w^k}{kk!(2k+1)}\right] \tag{4.72}$$

where $w = \alpha^2 r^2/4\Delta t$ and C is Euler's constant, $C = 0.52777\ldots$.

The other integrals do not have a singularity due to the orthogonality of r and n. Also the integrals for $s \neq m$ are free of singularity since this is cancelled due to the subtraction of $E_1(w_i)$ and $E_1(w_{i-1})$ (see eqn (4.67)).

However, the series appearing in the above formula converges very slowly, especially when the argument w is not small enough. The way to overcome this problem consists in choosing the domain of integration such that $w = 1$. In other words one has to take the small vicinity around the centre of the element, say, Δ_i to obtain $w = 1$. Then one obtains the following formula with 8-digit accuracy:

$$g_{mmii} = \frac{\alpha^2}{2\pi}\left[0.29024786\Delta_i + \int_{\Delta_i}^{e_i/2} E_1(w)dr\right] \tag{4.73}$$

The integral can be calculated numerically since it is deprived of singularity.

Also the discussion carried out for the sinusoidal case concerning the quasi-singularity remains valid here and for the case considered one needs to use as 'added–subtracted' functions

$$\ln w \text{ for matrix } \boldsymbol{G}$$

$$\frac{1}{r^2} \text{ for matrix } \boldsymbol{H}$$

Regardless of the methods adopted for solving the integrals in terms of time, the domain integrals occuring in the vector $\{F\}$ have to be calculated, assuming, of course, that $u_0(\boldsymbol{x}) \neq 0$ or $f \neq 0$. These calculations may sometimes cause some difficulties and in this case it would be advisable to avoid them. The way of calculating such an integral consists in substituting the domain integral by the boundary one. Namely we have:

$$I_0(\boldsymbol{x}) = \int_\Omega u_0(\boldsymbol{x})v(\boldsymbol{x};\boldsymbol{x}_i,t,\tau = 0)d\Omega \tag{4.74}$$

where v is given by eqn (3.90).

In the two-dimensional case one can substitute the above integral by the one-dimensional boundary integral using the Green's identity:

$$\int_\Omega(\nabla^2 uV - \nabla^2 Vu)d\Omega = \int_\Gamma\left(u_0\frac{\partial V}{\partial n} - \frac{\partial u_0}{\partial n}V\right)d\Gamma \tag{4.75}$$

Assuming that u_0 satisfies the Laplace's equation (that is, is a harmonic function in Ω) and taking V as a solution of the following Poisson's equation

$$\nabla^2 V = -v \tag{4.76}$$

one obtains

$$I_0 = \int_\Gamma \left(u_0 \frac{\partial V}{\partial n} - \frac{\partial u_0}{\partial n} V \right) d\Gamma \tag{4.77}$$

Of course, the problem is how to determine function V but since it can be any arbitrary solution of eqn (4.76) it can be found without special difficulty. For the two-dimensional case one such solution is:

$$V = \frac{1}{8\pi} Ei(\alpha^2 r^2/4t) \tag{4.78}$$

In practice, however, this problem may not exist as the initial condition may vanish or, when the magnetic vector potential is taken as a variable, the initial value of A is not a harmonic function. Nevertheless, in some cases the procedure presented above may be useful.

The same procedure can be employed to quasi-stationary cases when the non-homogeneous Helmholtz equation is considered. Function V is then as follows:

$$V = \frac{1}{2\pi\alpha^2} K_0(\alpha r) \tag{4.79}$$

The validity of the BEM in transient process analysis reveals itself when eddy currents are to be examined. It is most clearly seen in the two-dimensional case treated by the magnetic vector potential A. Then the eddy current density results from the time differentiation of A which is made numerically in the domain methods. Needless to say that it can and does introduce an additional error. The use of the BEM enables solving the problem without loss of accuracy since the differentiation in time can be made analytically. Indeed, due to an analytical form of integrands one can compute eddy currents directly, making use of some elements of the convolution theorem. The integrals occurring in eqn (4.71) are the convolution integrals with respect to time and then their time differential have the following forms:

$$\frac{d}{dt} \int_0^t f(\tau) g(t-\tau) d\tau = \int_0^t f(\tau) \frac{\partial}{\partial t} g(t-\tau) d\tau \tag{4.80}$$

Coming back to eqn (4.71) we have to calculate the time differentials of the functions β and φ and in special cases v. All these integrals are expressed by the following formulae, respectively:

$$\frac{\partial\beta}{\partial t} = \frac{\alpha^4}{8\pi}[\exp(-w_{i-1})/(t-t_{i-1})^2 - \exp(-w_i/(t-t_i)^2] \qquad (4.81)$$

$$\frac{\partial\varphi}{\partial t} = \frac{\alpha^2}{4\pi}[\exp(-w_{i-1})/(t-t_{i-1}) - \exp(-w_i)/(t-t_i)] \qquad (4.82)$$

$$\frac{\partial v}{\partial t} = \frac{\alpha^2}{4\pi(t-\tau)^2}\exp\left[\frac{\alpha^2 r^2}{4(t-\tau)}\right]\left[\frac{\alpha^2 r^2}{4(t-\tau)} - 1\right] \qquad (4.83)$$

Using the above formulae one can compute the eddy currents and hence the eddy current loss, forces and the other integral parameters. Also the boundary technique makes possible the computation of space derivatives in a similar way. From them one can obtain analytically field values (if the magnetic vector potential A is used) or the eddy current density, if the stream function u has been employed. It should be repeated once again that such possibilities are unapproachable by means of the other methods, especially those of the domain type. The semi-analytical way of eddy current calculation is of special importance when the skin effect has to be analysed. Then the analyst can choose the internal points as close to the boundary as he or she wishes and hence the analysis gives more accurate results near the skin than by using other methods which are strongly limited by using a regular mesh.

The BEM in transient problems has been illustrated by the sample calculations for test model (Fig. 4.14). It consists of a conductor in the slot of an electrical machine where the distribution of eddy current density is calculated (Fig. 4.14) as well as the integral values (resistance and reactance) (Fig. 4.15) (Krawczyk 1988). Eddy currents are induced here by switching on current of the unit step form.

Starting the description of the boundary element method in three-dimensional problems we shall come back to Section 3.2.4 in which the $\boldsymbol{H}-\phi$ approach using the boundary-integral way has been presented. This case can be also treated by the BEM and approximating the boundary S of the volume V by flat elements $\{S_i\}$ (polygons) and the boundary values by constant values with respect to space and time one obtains the following system of linear equations (Krawczyk and Wesełucha 1992):

$$\begin{aligned} 0.5\boldsymbol{j}_{s_k}(t) &+ \sum_{j=1}^{N}\boldsymbol{n}_k \times \sum_{i=1}^{m}\boldsymbol{j}_{s_j}(t_i)\times \boldsymbol{a}_{kj}(t,t_i,t_{i-1}) \\ &- \frac{1}{4\pi}\sum_{j=1}^{N}\sigma_j(t)\boldsymbol{n}_k \times \boldsymbol{b}_{kj}(t) \\ &= -\boldsymbol{n}_k \times \boldsymbol{H}_{e_k}(t) \end{aligned} \qquad (4.84)$$

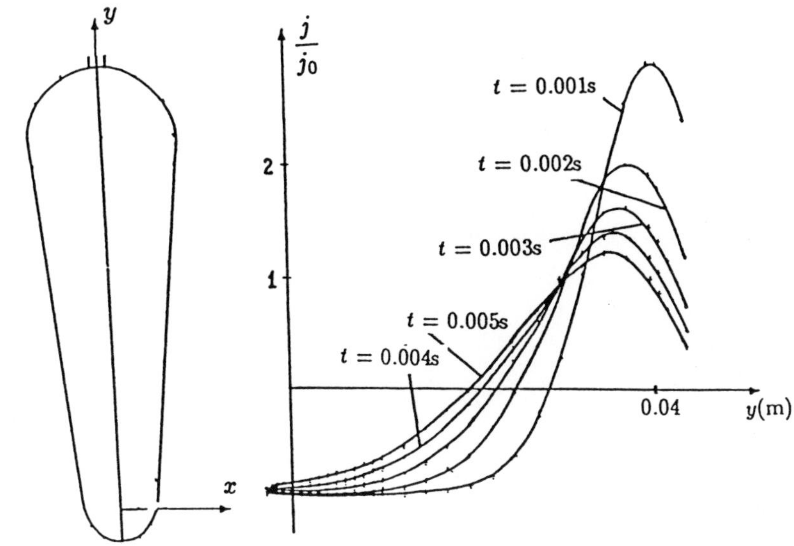

FIG. 4.14. Model considered and eddy current density distribution

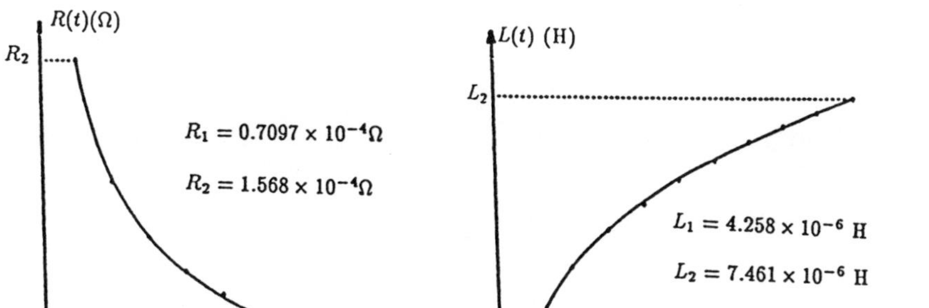

FIG. 4.15. Curves of resistance and inductance of conductor of Fig. 4.14

$$
\begin{aligned}
0.5\sigma_k(t) &+ \frac{1}{4\pi}\sum_{j=1}^{N}\sigma_j \boldsymbol{n}_k \cdot \boldsymbol{b}_{kj}(t) \\
&- \sum_{j=1}^{N}\boldsymbol{n}_k \cdot \sum_{i=1}^{m}\boldsymbol{j}_{s_j}(t_i) \times \boldsymbol{a}_{kj}(t, t_i, t_{i-1}) \\
&= -\boldsymbol{n}_k \cdot \boldsymbol{H}_{e_k}(t)
\end{aligned} \tag{4.85}
$$

where $k = 1 \ldots N$, m is the number of time steps and

$$\boldsymbol{a}_{kj} = \int_{S_j} \boldsymbol{n}_j \psi(r_{kj}, t, t_i, t_{i-1}) dS \tag{4.86}$$

$$\boldsymbol{b}_{kj} = \int_{S_j} \boldsymbol{n}_j \frac{1}{r_{kj}^2}\frac{\partial r}{\partial n} dS \tag{4.87}$$

$$\Psi(r_{MP}, t, t_i, t_{i-1}) = \frac{\alpha^2}{2\pi^{3/2} r_{MP}^3}[\Gamma(\tfrac{3}{2}, w_{i-1}) - \Gamma(\tfrac{3}{2}, w_i)] \tag{4.88}$$

$$w_i = \frac{\alpha^2 r_{MP}^2}{4(t - t_i)}$$

$\Gamma(p, x)$ is the incomplete Gamma function.

The features of function Ψ are shown in Section 3.2.4.

It seems that some comments are necessary as to the treatment of time. Namely, in the direct formulation (that of the two-dimensional case) two possibilities were discussed: one is the sequential way (step by step) and the second the so-called cumulative way. Here the second way is preferable as the determination of the initial state for each step seems to be difficult, even impossible.

As the problem is of vectorial nature there exists a need of arranging all the spatial vectors in order. Taking for simplicity the first time interval one can write the system of boundary-integral equations in the form:

$$0.5\alpha^2 \boldsymbol{j}_s(P) + \int_S [F_1]\boldsymbol{j}_s \Psi dS - \frac{1}{4\pi}\int_S \{G\}\sigma \frac{1}{r_{MP}^3} dS = -\boldsymbol{H}_t \tag{4.89}$$

$$0.5\sigma(P) + \int_S \{G\}\boldsymbol{j}_s^T \Psi dS + \frac{1}{4\pi}\int_S F_2 \sigma \frac{1}{r_{MP}^3} dS = -\mu_0 H_n \tag{4.90}$$

where the vectors and matrices result from the geometrical configuration and inner and outer products of normal and radial unit vectors. Denoting the space coordinates as $\{1,2,3\}$, the vectors and matrices are as follows:

$$[F_1] = \begin{bmatrix} n_3 r_3 + n_2 r_2 & -n_1 r_2 & -n_1 r_3 \\ -n_2 r_1 & n_1 r_1 + n_3 r_3 & -n_2 r_3 \\ -n_3 r_1 & -n_3 r_2 & n_2 r_2 + n_1 r_1 \end{bmatrix}$$

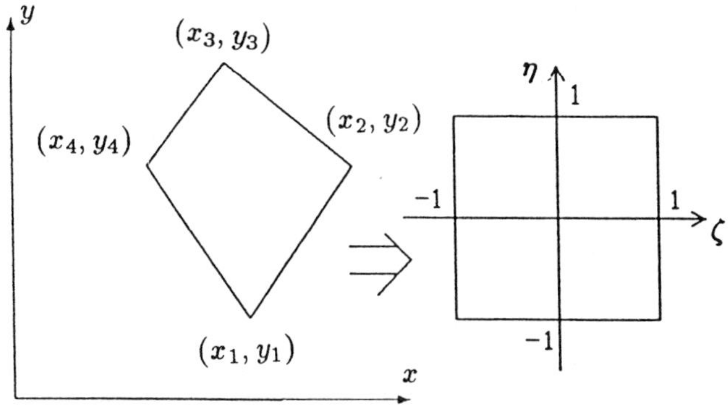

FIG. 4.16. The transformation of quadrangle

$$\{G\} = \left\{ \begin{array}{c} n_2 r_3 - n_3 r_2 \\ n_3 r_1 - n_1 r_3 \\ n_1 r_2 - n_2 r_1 \end{array} \right\}$$

where n_1, n_2, n_3 are the components of the unit normal vector $\boldsymbol{n}$ at point P and r_1, r_2, r_3 are the components of vector $\boldsymbol{r}_{MP}$ and are as follows:

$$r_1 = r_{MP} \cos\beta \cos\varphi, r_2 = r_{MP} \sin\beta \cos\varphi, r_3 = r_{MP} \cos\varphi$$

The angles β and φ may be easily evaluated from the coordinates of the vector $\boldsymbol{r}_{MP}$. Function F_2 is the simple inner product of vectors $\boldsymbol{n}_P$ and $\boldsymbol{r}_{MP}$.

To establish the boundary integrals for $k \neq j$ one can use the quadrature technique and taking the appropriate number of nodes one can expect satisfactory results. Thus, we have

$$\int_S f(x,y)dS = \sum_{i,j}^{n_Q} w_i w_j f(\zeta(x,y), \eta(x,y)) \text{abs}(J) \qquad (4.91)$$

where w_i, w_j are the weights of quadrature, ζ, η are the coordinates after transformation of quadrangular element to a standard square (-1,1), J is the Jacobian of transformation (Fig. 4.16). The accuracy of integration is determined by the order of quadrature, that is, by n_Q. In order to reduce the number of nodes a special criterion is introduced: the longer the distance between elements the smaller the number of nodes. This reduces the time of calculation to a great extent.

In the case of $k = j$ the numerical integration fails since the integrands involve the singularity of $1/r^n$ type. Then special procedures should be adopted and the one most often used is presented below. The integrals (only existing if $k = j$) can be transformed to the forms:

$$I_1 = \int_{S_j} \boldsymbol{n}_j \times \frac{\boldsymbol{r}_{jj}}{r_{jj}^3} dS = \int_{S_j} \boldsymbol{n}_j \times \operatorname{grad} \frac{1}{r} dS \tag{4.92}$$

$$I_2 = \int_{S_j} \boldsymbol{n}_j \cdot \boldsymbol{j}_s \times \boldsymbol{r}_{jj} \Psi dS = \int_{S_j} \boldsymbol{n}_j \cdot \operatorname{curl}(\boldsymbol{j}_s K) dS \tag{4.93}$$

Adopting vector analysis theorems gives:

$$I_1 = \int_{l(S_j)} \frac{1}{r_{jj}} dl \tag{4.94}$$

$$I_2 = \int_{l(S_j)} \boldsymbol{j}_s K dl \tag{4.95}$$

where K is given by eqn (3.103) and $l(S_j)$ is the boundary of element S_j.

This way the surface singular integrals are substituted by line integrals which are not singular. Then the numerical integration can be safely adopted.

Since the system of boundary-integral equations enables us to obtain the fictitious surface values only, it is necessary to compute the internal values: magnetic field strength and eddy current density which are the most interesting. Thus, having calculated $\boldsymbol{j}_s$ one can write:

$$\boldsymbol{H}^+(P,t) = \sum_{j=1}^{N} \sum_{i=1}^{n} \boldsymbol{j}_{sj}(t_i) \times \int_{e_j} \boldsymbol{r}_{Pj} \Psi(M,P,t,t_i,t_{i-1}) dS \tag{4.96}$$

The eddy current density is calculated from the fundamental rule $\boldsymbol{j} = \operatorname{curl} \boldsymbol{H}^+$. Making use of vector identities after some rearrangements one obtains:

$$\begin{aligned} \boldsymbol{j}(P,t) = & \int_S \sum_{i=1}^{n} \boldsymbol{j}_s(M,t_i), [K(P,M,t,t_{i-1}) - K(P,M,t,t_i)] dS \\ & + \int_S \sum_{i=1}^{n} \operatorname{grad}(\boldsymbol{j}_s(M,t_i) \cdot \boldsymbol{r}_{MP} \psi(P,M,t,t_{i-1},t_i)) dS \end{aligned} \tag{4.97}$$

The magnetic field in the exterior region can be established by means of eqn (3.104) and basic relation $\boldsymbol{H}^- = \operatorname{grad} \phi^-$. Thus, finally one gets

$$\boldsymbol{H}^-(P,t) = \boldsymbol{H}_e(P) - \frac{1}{4\pi} \sum_{j=1}^{N} \sigma_j(t) \int_{e_j} \boldsymbol{r}_{Pj} \frac{1}{r_{Pj}^3} dS \tag{4.98}$$

As it was already mentioned in Chapter 3 the methodology of solving the problem with sinusoidal time dependence remains the same. The change consists in taking the kernel of eqn (3.102) appropriate for Helmholtz's equation and this is expressed by

$$K(\boldsymbol{r}) = \frac{\exp(-i k \boldsymbol{r})}{4\pi \boldsymbol{r}} \tag{4.99}$$

It seems that more time was devoted to the boundary element technique than to the other previously discussed methods. However, it appears that this method, or rather its use grows continuously throughout the electromagnetic world.

Application of the BEM

The BEM is the method that has gained great importance and attracted the researchers' interest since the late 1970s. Its application in electrical engineering is rather new because at the beginning BEM activity was devoted to the development of the method and not to its technical use. Considerably more technical applications of the BEM surfaced more recently.

It is very easy to find the illustrative examples from the proceedings of international conferences such as COMPUMAG, INTERMAG, ICEM, ISEM, ISEF and the like. One example has already been given in the previous section. There are several research groups which deal with the BEM in electrical engineering. One can mention here the following references in modelling of eddy currents by BEM: Salon and Schneider (1981), Kost (1989), Rucker and Richter (1983), Krawczyk (1990), Krajewski and Salon (1991), Enokizono (1991), Tsuboi and Misaki (1987). The list of references is indicative only and far from complete. Indeed, many scientific centres deal with the method.

However, in spite of this fact, there are not as many commercial packages as for FEM. It is obvious that particular research centres have developed some BEM softwares for their own usage (Rucker and Richter 1990) but their availability is not as wide as those of FEM. The reason for this is that BEM is still being developed and many papers from international conferences consider just pure methodological problems, for example, integration of singularity.

To illustrate how the method works the technical application concerning the analysis of the end zone of turbogenerator is shown. The model analysed is described in Fig. 4.17.

Krajewski and Salon (1991) calculated the eddy current density in the pressure plate by means of BEM and compared with FEM. In Figs 4.18 and 4.19 the distributions of eddy current density are presented. One can see that BEM and FEM give compatible results. It is interesting to see that the authors recommend neither BEM nor FEM, but leave the final choice to the user. Readers of this book should not be surprised at such a conclusion since it is our deep, constantly repeated belief that the evaluation of methods can, by no means, be made *a priori*.

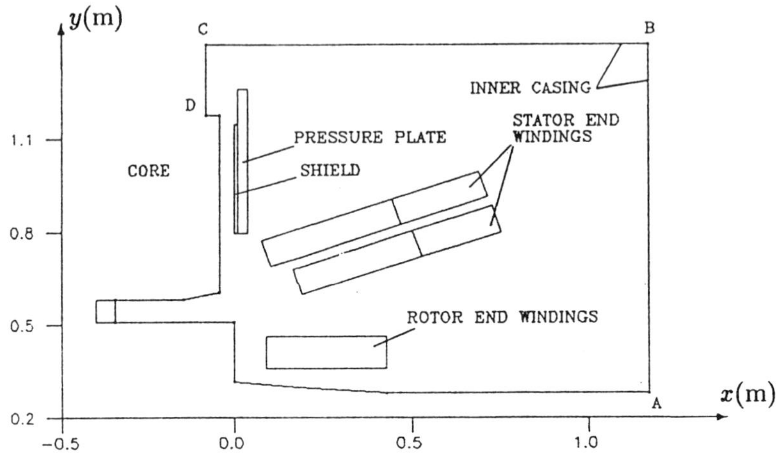

FIG. 4.17. Model of the end zone of turbogenerator

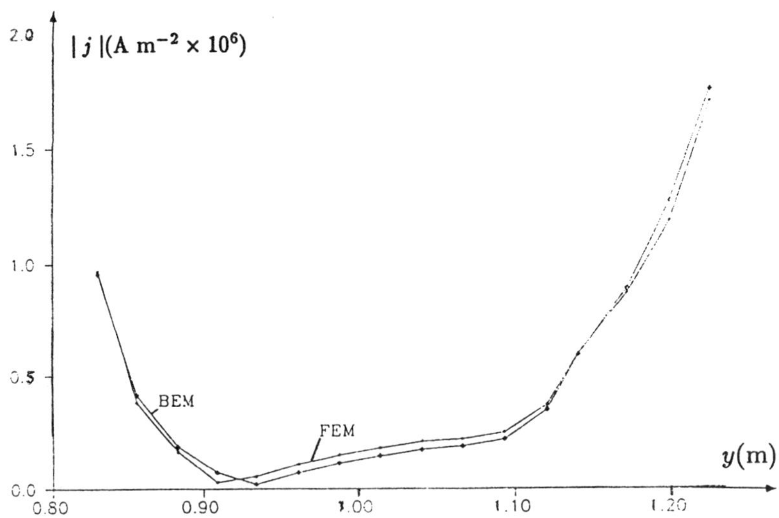

FIG. 4.18. Magnitude of the eddy current density on the middle line of the pressure plate

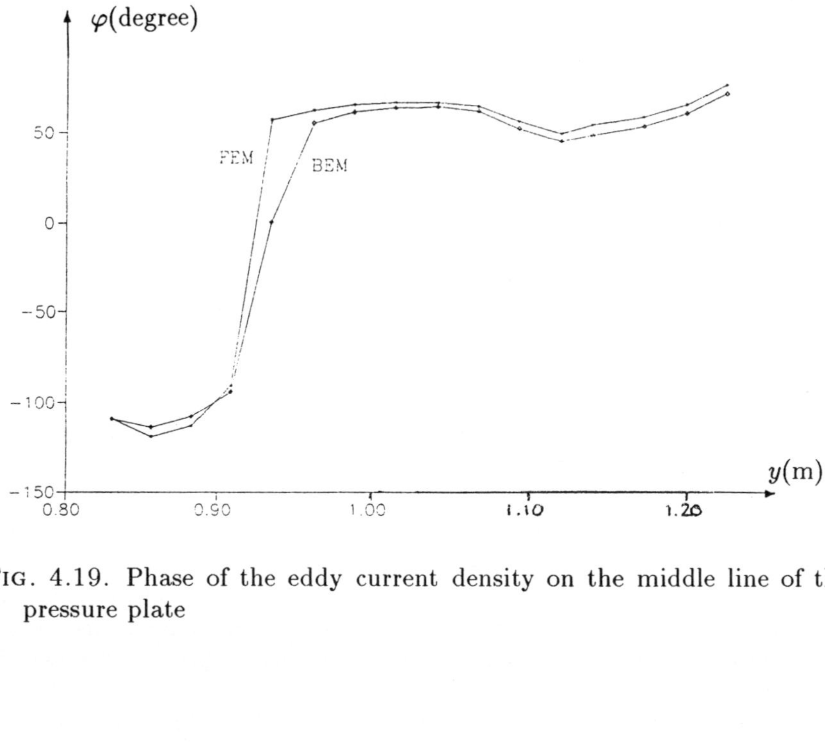

FIG. 4.19. Phase of the eddy current density on the middle line of the pressure plate

5

TREATMENT OF SPECIAL CASES

All chapters written up to this point have an *implicit* assumption, namely that configurations considered include only a homogeneous region. In practice, though, we deal with at least two cases which are not of this type. The first one is the problem when the eddy current region is surrounded by non-conducting infinite space (open boundary problems) and the second one is where within the conductor different material parameters exist (interface problem). The impatient reader also ask about nonlinearity, but as the subject is specific and rather wide it will be presented in Chapter 6. Thus let us first discuss the problems mentioned above.

5.1 Open boundary problems

The history of the open boundary problem is strongly connected with the development of numerical techniques. Before the computer era many analytical methods and formulae did not consider the fact that a conductor is usually placed in a free-space and hence there are no boundary conditions prescribed except those at infinity. Also, the boundary conditions are prescribed for simple geometries, for example, long conductor with circular cross-section. But reality is such that conductors are located in the non-conducting space (usually in air) so that the boundary conditions on the conductor surface are unknown. The classical approach in this case is to consider two regions in which the field is to be solved and use the well-known compatibility (or interface) conditions in order to establish unknown boundary values. Of course, such a procedure can be adopted also in the pure analytical approach; however, it would be very difficult to proceed to a solution.

Initially also numerical approaches avoided the fact of non-homogeneity and this problem surfaced in the mid-1970s, when the term 'open boundary problem' was introduced. The main difficulty of such problems is that we do not have boundary conditions and instead of them we have to put the interface conditions to the solution. Thus, for both the interior space (conducting) and the exterior space (non-conducting) there are two unknown vectors: one is the field vector and the second is the boundary vector.

The classical FEM approach uses the fact that the conditions at infinity are always given, namely, the conducting region was surrounded by an artificial line being placed far enough away to simulate infinity. Then on this line the conditions at infinity (homogeneous Neumann type) are prescribed and the problem becomes more tractable, having a material non-homogeneity. It will be shown in the next section that the FEM is just the proper method for analysing such models. After the solution of the extended region (with rather large elements) one obtains the boundary values of a function and then the process is repeated for the conducting region. The problem is, however, where the artificial line should be drawn to ensure that eddy currents will not influence the field. On the other hand the domain cannot be freely enlarged since the dimension of the problem (in the sense of linear algebra) should not be too big. Compromises in this respect are possible but generalization does not seem feasible. From this point of view the approach finds little interest among researchers. They look for other approaches which use mixed methods.

One of this approaches is the method connecting the FEM and the analytical solutions in the exterior space. It is a very fruitful approach and much more accurate than that discussed above. The analytical solution is usually, if not always, given in the form of Fourier's series which limits somehow the scope of possible applications: conductors should be geometrically regular (rectangular or circular). This idea is developed in a series of papers (for example, Gasiorski 1985).

The next step approach is the joint application of the FEM and the BEM. Here one uses the fact that the BEM can be employed either for exterior or interior regions. Thus, using the FEM in solving the Helmholtz (or parabolic) equation and the BEM in solving the Laplace (Poisson) equation and coupling the solutions on the boundary, one gets a system of two matrix equations with two unknown vectors. As none of the approaches is entirely satisfactory the approach of coupling FEM and BEM has an essential disadvantage: that the final matrix is non-symmetrical and so one of the fundamental FEM advantages vanishes. This drawback results from the nature of the open boundary problem itself.

The next possibility is using the so-called 'infinite elements'. Briefly speaking, the basis functions have a form which maps the behaviour of a searched function in the infinite space as it really is. This approach is of wide use and can be supposed to fit the open boundary problems. The term 'infinite element' suggests that it really covers the infinite space. It is therefore useful to describe the approach in more detail.

Consider two regions: the first one Ω_i that is finite and conducting and the second one Ω_e that is of infinite extent. The first region is subdivided as usual in the finite element method while the second one is covered by the infinite elements, that is, the elements having one side determined (boundary segment) and two other sides tending to infinity (Fig. 5.1).

The two regions should satisfy typical interface conditions. The behaviour of the magnetic field at infinity is expressed by the basis (shape) function which determines how the field decays. Almost all researchers

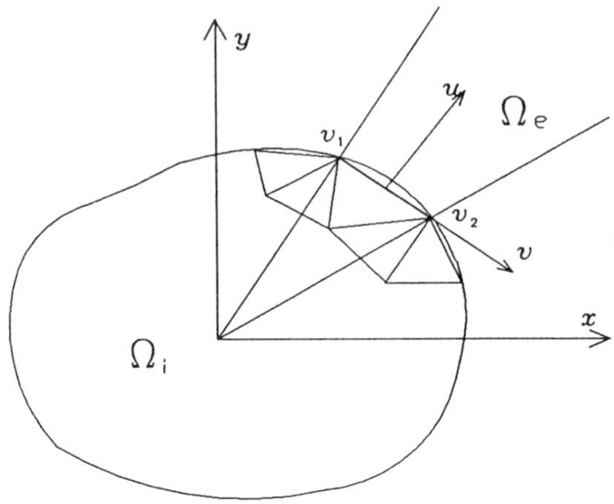

FIG. 5.1. Infinite element

suggest that the basis function should be given as a power of inverse radius. Introducing a new system of coordinates $0uv$ (Fig. 5.1) one obtains the following approximation on $(i, i+1)$ element:

$$u(v_i, v_{i+1}) = u_i\varphi_i + u_{i+1}\varphi_{i+1} \tag{5.1}$$

where $\varphi_i = (a_i v + b_i) f(u)$ and $f(u)$ is, in turn, the *decay* function.

The coefficients a_i and b_i represent the linear approximation along the $(i, i+1)$ element and result from the classical one-dimensional finite element technique. Then, in order to find the matrix element, one has to calculate the following integrals:

$$\int_{v_1}^{v_2}\int_{u_0}^{\infty}\left(\frac{\partial\varphi_i}{\partial u}\frac{\partial\varphi_j}{\partial u} + \frac{\partial\varphi_i}{\partial v}\frac{\partial\varphi_j}{\partial v}\right) du dv \tag{5.2}$$

The above integrals are representatives of the infinite region in the finite element matrix. The problem arises, however, in choosing the *decay* function $f(u)$. This function is mainly, as mentioned above, a power of inverse radius. In practice the first power r^{-1} is used. Some attempts have been made to take a function that would be more complex (Nakata *et al.* 1988) but eventually all of them accepted the form of inverse radius. It seems that the infinite element technique is one of the most attractive proposals in solving the open boundary problems.

If joining FEM and BEM does not give satisfactory results, it may be, perhaps, that using pure BEM will bring what is really expected. For this reason the approach will be discussed more in detail. First, let us determine

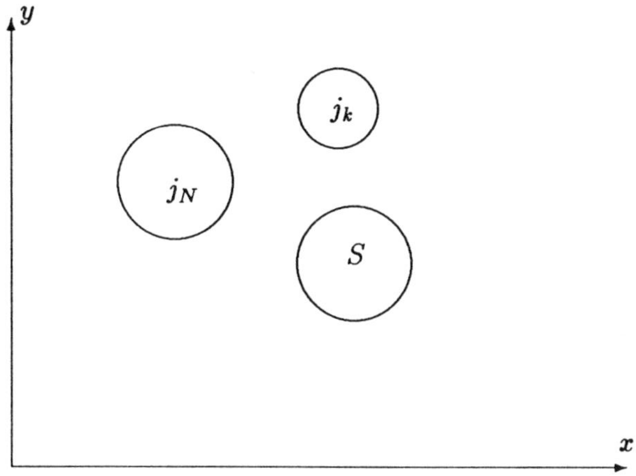

FIG. 5.2. Two-dimensional system of conductors

the object analyzed. Consider a two-dimensional domain Ω placed in the Cartesian coordinates (Fig. 5.2). Assume that the excitation is external, that is, the set of exciting conductors is located outside the domain Ω. The material parameters of the conducting region are μ and γ. Since the problem is two-dimensional the magnetic vector potential is to be applied here. Then the whole process is described by the following set of equations:

$$\nabla^2 A = \mu\gamma \frac{\partial A}{\partial t} - \mu \operatorname{grad} \phi \qquad \text{in } \Omega \tag{5.3}$$

$$\nabla^2 A = -\mu_0 \sum_{k=1}^{N} j_k(x, y, t) \qquad \text{in } R^2 \setminus \Omega \tag{5.4}$$

The term grad ϕ represents the electric field connected with the flow of eddy currents and is constant in the two-dimensional $A - \phi$ approach (see Chapter 3); hence, it will be denoted as C. Both equations can be solved by means of BEM and then the interface conditions on the boundary should be taken into account. One can prove that BEM is also effective for an exterior Laplace problem (Brebbia *et al.* 1984).

In the case of Laplace's equation the fundamental solution is of logarithmic type. The procedure of obtaining boundary-integral form for static cases is well-known so that is not described here. What is more, there are no essential differences, except that of the form of the fundamental solutions, between boundary-integral form for Laplace's and Hemholtz's equations.

The discrete forms of the two equations for the transient phenomenon are as follows:

$$[H_L]\{A\} - [G_L]\{B_t\} = \{F_L(t)\} \tag{5.5}$$

$$[H(t)]\{A(t)\} - [G(t)]\{B_t(t)\} = \{F(t)\} \tag{5.6}$$

where B_t is the tangential component of magnetic flux density.

Then introducing the conditions that hold on the eddy current region boundary

$$A_1 = A_2 \tag{5.7}$$

$$\frac{1}{\mu_1}\frac{\partial A}{\partial n}|_1 = \frac{1}{\mu_2}\frac{\partial A}{\partial n}|_2 \tag{5.8}$$

one obtains the matrix in the form of:

$$\begin{bmatrix} \boldsymbol{H}(t) & -\boldsymbol{G}(t) & H_0(t) \\ -\boldsymbol{H}_L & \boldsymbol{G}_L & 0 \end{bmatrix} \begin{Bmatrix} \boldsymbol{A}(t) \\ \boldsymbol{B}_t(t) \\ C(t) \end{Bmatrix} = \begin{Bmatrix} \boldsymbol{F}(t) \\ \boldsymbol{F}_L(t) \end{Bmatrix} \tag{5.9}$$

Note that because of the electric field the system is undetermined and needs additional equations.

At this place, however, one should introduce the so called 'constraints', which give information on the eddy current flow. The constraint condition is expressed in integral form and depends on the way in which the conductors are connected at infinity or whether they are active or passive. From these points of view one can consider the following states:

(1) all conductors are independent of each other;
(2) all conductors are connected in series;
(3) all conductors are connected in parallel.

As to state 1 the constraint condition tells us about the total current in each conductor, hence we have as many conditions as the number of conductors. Assuming for simplicity that eddy current flows in the k-th conductor and the external current flowing through this conductor is equal to I_k one can write using Ampere's law:

$$\int_{\Gamma_k} B_t(x, y, t)d\Gamma = \mu I_k(t) \tag{5.10}$$

$k = 1 \ldots N$ (number of conductors).

This integral can be written in the discrete form as follows:

$$\sum_{i=1}^{n_k} B_{t_i} N_i = \mu I_k(t) \tag{5.11}$$

where n_k is the number of boundary elements of the boundary Γ_k, N-shape function (in constant approximation $N_i = e_i$), or in matrix form:

$$\{N\}^T \cdot \{B_t(t)\} = -\mu I(t) \tag{5.12}$$

Finally, the matrix equation with the constraint condition will be as follows:

$$\begin{bmatrix} \boldsymbol{H}(t) & -\boldsymbol{G}(t) & H_0(t) \\ -\boldsymbol{H}_L & \boldsymbol{G}_L & 0 \\ \boldsymbol{0} & \boldsymbol{N} & 0 \end{bmatrix} \begin{Bmatrix} \boldsymbol{A}(t) \\ \boldsymbol{B}_t(t) \\ C(t) \end{Bmatrix} = \begin{Bmatrix} \boldsymbol{F}(t) \\ \boldsymbol{F}_L(t) \\ -\mu I(t) \end{Bmatrix} \tag{5.13}$$

As to state 2, it is obvious that the same current will flow in each conductor so that we have to impose only one condition instead of N as was in state 1. Thus, only one equation will be added to the the matrix equation (5.9).

As to state 3, the situation is similar to the first one; however, the domain of each integration will obey now the sum of surfaces of the conductors connected in parallel while the current will be the total current flowing in the set of the conductors.

This method of analysis seems to overcome the problem of grad ϕ mentioned earlier, as it does not require any pre-initial consideration as to whether grad ϕ exists or not. Moreover, since the BEM matrix is full, the addition of some equations does not destroy its structure as is the case with the FEM matrix.

So far the problem with current excitation was discussed. In many technical cases we face the problems with flux excitation, that is, eddy currents are induced due to a magnetic field coming from beyond the region of intrest. If the exciting field is homogeneous there is no problem in finding a solution by using the above considerations *mutatis mutandis*. It is easy to notice that the integral of exciting currents establishes a magnetic vector potential (Newtonian potentials). Using it one can easily calculate the magnetic vector potential from a given field B. Let us say that the field is of y-direction. Then the part of right-hand side vector of eqn (5.13) is

$$\boldsymbol{F}_L = B(x - x_0) \tag{5.14}$$

where x_0 is the reference point.

The three-dimensional problems are usually considered in the open boundary mode and do not require any special treatment. They are described by a pair of quantities joining exterior and interior regions (see Section 3.2).

It should be stressed here that the BEM analysis of the open boundary problems does not require any artificial condition and is mathematically correct and unique.

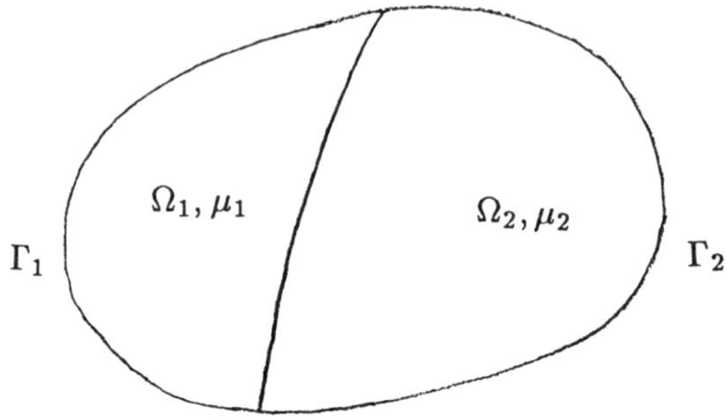

FIG. 5.3. Interface simple-connected problem

5.2 Interface problems

In many technical applications one meets the situation when the region of interest contains materials of different conductivities or permeabilities. Then this region can be named a piecewise homogeneous region and depending on the method being used it requires some special treatment. For this treatment the domain methods may be used which actually by their nature are well-suited to such problems. This results from the fact that both the FDM and FEM act on the discretized regions, thus in an extreme case each element can have different material parameters. What is more, the variational or 'weak' mathematical formulation includes the interface conditions in itself, if one deals with the magnetic vector potential and for this reason it is preferred for two-dimensional problems. Indeed, considering a region consisting of two sub-regions (Fig. 5.3) and writing the 'weak' formulation for both sub-regions it can be easily proved that the surface integrals on the interface cancel each other, if the classical interface conditions are introduced, namely:

$$A_1 = A_2 \tag{5.15}$$

$$\frac{1}{\mu_1}\frac{\partial A}{\partial n}\Big|_1 = \frac{1}{\mu_2}\frac{\partial A}{\partial n}\Big|_2 \tag{5.16}$$

Then considering the region Ω consisting of two subregions: Ω_1 and Ω_2 one can write the weak formulation for both regions:

$$\int_{\Omega_1}\left(\frac{1}{\mu_1}\operatorname{grad} A \cdot \operatorname{grad} v - \beta^2 A v\right) d\Omega = \int_{\Gamma_1}\frac{1}{\mu_1}\frac{\partial A}{\partial n} v\, d\Gamma + \int_{\Gamma_i}\frac{1}{\mu_1}\frac{\partial A}{\partial n} v\, d\Gamma \tag{5.17}$$

$$\int_{\Omega_2}\left(\frac{1}{\mu_2}\operatorname{grad}A\cdot\operatorname{grad}v-\beta^2Av\right)d\Omega=\int_{\Gamma_2}\frac{1}{\mu_2}\frac{\partial A}{\partial n}vd\Gamma-\int_{\Gamma_i}\frac{1}{\mu_2}\frac{\partial A}{\partial n}vd\Gamma \tag{5.18}$$

where $\beta^2 = i\omega\gamma$.

Adding both equations one can easily notice that both boundary integrals over Γ_i cancel each other due to eqn (5.16). In addition, the approximation technique, which is of interpolation type, obeys automatically the interface conditions. Finally, one can write one equation treating the problem as homogeneous and putting only the proper values of μ at the elements.

The FDM in interface problems is more difficult because one has to build a finite difference grid near the interface or just include the interface point.

Consider the Helmholtz equation

$$\nabla^2 A = \alpha^2 A \tag{5.19}$$

which holds in two materially different regions, say, with different permeabilities.

Its difference scheme, assuming that the central point lies on the interface line is difficult to describe as one cannot determine the material parameter α there. There exist various ways of analyszing such a case and the simplest one is based on the concept of a harmonic mean value. However, one should remember that the interface conditions are to be ensured. It is usually done by enforcing additional relations linking nodal values of both sub-regions. The way one should proceed then is rather cumbersome so that the variant of FDM derived from a differential equation is of small usefulness. However, there is also the variant of FDM based on integral form (see eqn (4.22)) and using this approach the interface problem can be naturally solved. Nevertheless, one must say that FDM is not, in the nature of things, dedicated to non-homogeneous (in the sense of material parameters) cases. Neither is BEM and serious problems appear when dealing with the boundary-integral approaches. Since these methods observe the boundary only we have to introduce additional information about the interior region to the boundary equation. Let us consider for simplicity the two subregions having the common boundary (Fig. 5.3).

Let boundary conditions be prescribed on the whole boundary Γ. On the interface, the interface conditions are to be fulfilled. Then for both subregions '1' and '2' one can write the boundary integral equations, which are as follows:

$$c_{i1}A_1+\int_{\Gamma_1}A\frac{\partial v}{\partial n}d\Gamma+\int_{\Gamma_i}A\frac{\partial v}{\partial n}d\Gamma$$

$$= \int_{\Gamma_1} v \frac{\partial A}{\partial n} d\Gamma + \int_{\Gamma_i} v \frac{\partial A}{\partial n} d\Gamma \tag{5.20}$$

$$c_{i2}A_2 + \int_{\Gamma_2} A \frac{\partial v}{\partial n} d\Gamma + \int_{\Gamma_i} A \frac{\partial v}{\partial n} d\Gamma$$
$$= \int_{\Gamma_2} v \frac{\partial A}{\partial n} d\Gamma + \int_{\Gamma_i} v \frac{\partial A}{\partial n} d\Gamma \tag{5.21}$$

After discretization they may be written in matrix form as follows:

$$[\boldsymbol{H}_{11}]\{\boldsymbol{A}_1\} - [\boldsymbol{G}_{1i}]\{\boldsymbol{B}_{ti}\} + [\boldsymbol{H}_{1i}]\{\boldsymbol{A}_i\} = [\boldsymbol{G}_{11}]\{\boldsymbol{B}_{t1}\} \tag{5.22}$$

$$[\boldsymbol{H}_{22}]\{\boldsymbol{A}_2\} - [\boldsymbol{G}_{2i}]\{\boldsymbol{B}_{ti}\} + [\boldsymbol{H}_{2i}]\{\boldsymbol{A}_i\} = [\boldsymbol{G}_{22}]\{\boldsymbol{B}_{t2}\} \tag{5.23}$$

Now taking the interface conditions one can get the one matrix equation in the form of:

$$\begin{bmatrix} \boldsymbol{H}_{11} & \boldsymbol{H}_{1i} & -\boldsymbol{G}_{1i} & 0 \\ 0 & \boldsymbol{H}_{2i} & \alpha\boldsymbol{G}_{2i} & \boldsymbol{H}_{22} \end{bmatrix} \begin{Bmatrix} \boldsymbol{A}_1 \\ \boldsymbol{A}_i \\ \boldsymbol{B}_{ti} \\ \boldsymbol{A}_2 \end{Bmatrix} = \begin{bmatrix} \boldsymbol{G}_{11} & 0 \\ 0 & \boldsymbol{G}_{22} \end{bmatrix} \begin{Bmatrix} \boldsymbol{B}_{t1} \\ \boldsymbol{B}_{t2} \end{Bmatrix} \tag{5.24}$$

As one can see the problem has expanded considerably, namely the number of additional equations is now $2N_i$ (N_i number of interface elements). This is a disadvantage of the boundary element method. However, this seems to be not so important if the number of sub-regions is relatively small. However, even if the number is large one can notice that in the final matrix there are a lot of zero blocks which make the process of solving the matrix equation faster and simpler. The case considered does not exhaust all possibilities, since the region is simply-connected. If the region is multi-connected the final matrix is slightly different and for a doubly-connected region it is as follows:

$$\begin{bmatrix} \boldsymbol{H}_{11} & \boldsymbol{H}_{12} & -\boldsymbol{G}_{12} \\ 0 & \boldsymbol{H}_{22} & -\alpha\boldsymbol{G}_{22} \end{bmatrix} \begin{Bmatrix} \boldsymbol{A}_1 \\ \boldsymbol{A}_2 \\ \boldsymbol{B}_{t2} \end{Bmatrix} = \begin{bmatrix} \boldsymbol{G}_{11} & 0 \\ 0 & \boldsymbol{I} \end{bmatrix} \begin{Bmatrix} \boldsymbol{B}_{t1} \\ \boldsymbol{F}(t) \end{Bmatrix} \tag{5.25}$$

Since the interface problems in the light of the FEM do not change anything in the general usage of the FEM, sample calculations are not necessary. Contrary to it the BEM for such cases should be considerably modified as compared to the standard one and therefore some sample calculations will be given.

Let us consider a doubly-connected region containing two sub-regions with different material parameters, say, γ_1 =0 and γ_2 = 56 MS/m (Fig. 5.4).

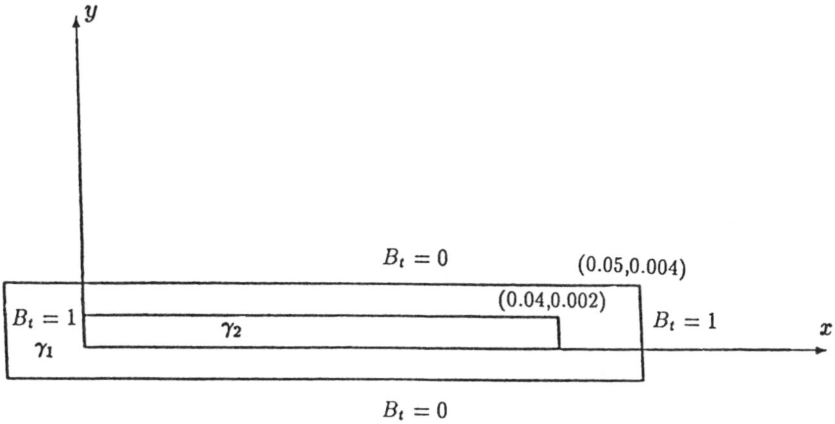

FIG. 5.4. Doubly-connected region

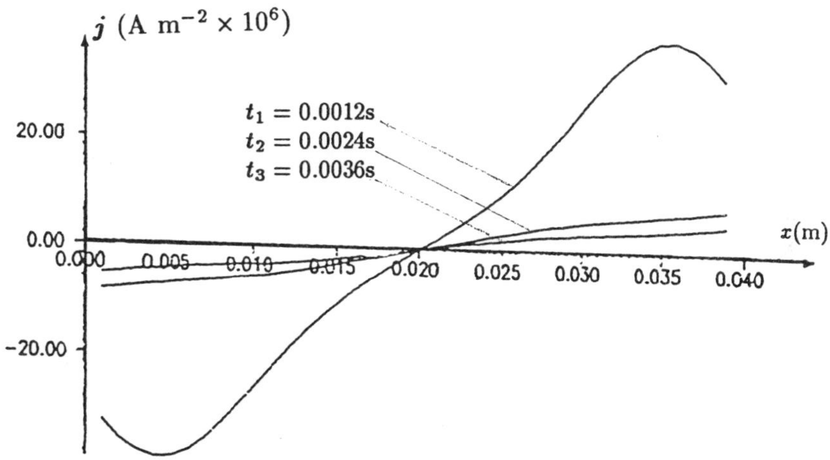

FIG. 5.5. Eddy current distributions

Assume the boundary conditions to be: Neumann condition ($\partial A/\partial n = 1(t)$) on parts of $x =$ const and homogeneous Neumann condition on parts $y =$ const. The case corresponds in some way to the open boundary case. Then the eddy current distributions at some time instants will be as shown in Fig. 5.5.

It is obvious that for problems containing more materially different subregions the BEM approach loses its advantages since the final matrix becomes larger and larger and comparable to the FEM matrix. The final choice of the method is left to the analyst whose preferences and experience are most important.

6

TREATMENT OF NON-LINEARITIES

Although eddy currents mainly exist in conducting materials which are by nature linear in the sense of constant (independent of field) permeability, they can be also analysed in non-linear materials. Indeed, it is quite common for eddy currents to flow in ferromagnetic media, in which case an analysis of their distribution and other integral parameters (power losses, forces etc.) is needed. It should be clearly stated that the problem of non-linearities is much more serious when eddy currents are present than in the static problems. It is especially seen in transient problems where the permeability is dependent on time. What is more, the hysteresis loop has to be then specified for each time-space point. For this reason the hysteresis will not be discussed here. As already mentioned (see Chapter 2) an extensive survey of approaches regarding non-linear materials is given by Stoll (1974).

6.1 Modelling of nonlinearity

It should be noticed that the entire mathematical representation of magnetic properties of ferromagnetic materials is almost impossible, especially regarding eddy currents. The dependence between $\boldsymbol{B}$ and $\boldsymbol{H}$ is expressed in two ways:

$$\boldsymbol{B} = \mu(H)\boldsymbol{H} \tag{6.1}$$

$$\boldsymbol{B} = \mu_0(\boldsymbol{H} + \boldsymbol{M}(H)) \tag{6.2}$$

where μ in the first formula is equal to $\mu_0\mu_r(H)$ while the vector $\boldsymbol{M}(\boldsymbol{H})$ in the second formula is the magnetization vector. These two forms are interchangable as to their physical meaning but have different mathematical properties which makes them useful for different cases. We shall deal here with the single-valued characteristics. The history of such a modelling is very long and is strongly connected with mathematical electromagnetism. Let us leave aside the approximation used for simple one-dimensional calculations, such as that by unit step function (see Stoll 1974). Looking for the method of approximation a compromise should be found between quality of approximation (accuracy) and its simplicity. The first, numerical attempts considered the polynomial interpolation, which can be shown on the example of Widger (1964):

$$\boldsymbol{B} = \frac{a_0 + a_1 H + a_2 H^2 + \cdots + a_n H^n}{1 + b_1 H + b_2 H^2 + \cdots + b_n H^n} \boldsymbol{H} \tag{6.3}$$

where $a_1, \ldots, b_1, \ldots$ are calculated from the experimental B–H curve or given commercially. Usually $n = 2$ and this produces a quite satisfactory approximation. Particularly taking into account the big discrepancy of material data (experimental and commercial) one should not pay great attention to the quality of approximation.

El-Sherbiny (1974) proposes another kind of interpolation, namely that of exponential nature. The series which models the B–H curve is of the form

$$B = a_0 + a_1 \exp(-b_1 H) + \cdots + a_n \exp(-b_n H) \tag{6.4}$$

It is suggested that taking $n = 4$ gives satisfactory representation.

The other possibility is the approximation given by Fröhlich and is as follows:

$$\boldsymbol{B} = \frac{\boldsymbol{H}}{a + bH} \tag{6.5}$$

Coefficients a and b result from the B–H curve. There exist a very wide set of possible approximations and it is not our aim to present all of them. One can mention here the approximations by hyperbolic functions, exponential functions, spline functions and the like. It seems that for technical goals the polynomial or Fröhlich approaches are quite satisfactory.

6.2 Numerical treatment

The problem of the analysis of eddy currents in non-linear media is a little more complex than other electromagnetic problems. For two-dimensional cases it is almost a rule that the magnetic vector potential is used. In this approach there is no simple relation between μ and A since μ is a function of B. Using any classical iterative methods one has to take a new value of μ knowing the value of H. But H in this case is given indirectly, namely as a derivative of A. Considering the numerical methods given in Chapter 4 one should notice that introducing non-linearity is different for each of them. It is relatively simple when dealing with the so-called 'domain methods' while for 'boundary methods' some difficulties appear. Let us discuss the procedures for these methods.

6.2.1 *Domain methods (finite difference method and finite element method)*

Introducing non-linearity is similar for the two methods. Since, for both, the domain is analysed there at internal points (elements) there is no difficulty in using the iterative method which uses the approximated B–H curve to improve, at every iteration, the value of μ as well as the magnitude searched for. It is easy to notice that the solution vector (H and/or A) is affected by non-linearity. For the finite difference method the procedure is extensively described in the book by Stoll (1974). In this book

Fröhlich's curve is used which seems to be the most appropriate and convenient. However, the problem may appear when the transient processes are analysed. Then the stability of solution may be difficult to ensure, which is inherently connected with the method itself. The same approach can be employed for the finite element method since the nodes are also internal.

For example, one can quote here the finite difference scheme given by Stoll. Consider one-dimensional problem described in terms of $\boldsymbol{H}$. Then the equation is as follows:

$$\frac{\partial^2 \boldsymbol{H}}{\partial x^2} = \gamma \frac{\partial B}{\partial H} \frac{\partial \boldsymbol{H}}{\partial t} \tag{6.6}$$

Using Frölich's curve, one obtains

$$\beta = \frac{\partial B}{\partial H} = \frac{a}{(a + bH)^2} \tag{6.7}$$

As long as the Crank–Nicholson scheme is used (see Section 4.1) the problem arises as to which time step should be taken into account to find the proper value of β. If one assumes that β does not vary very rapidly then one can take the value of β at the previous step for calculating the next one. This way small time steps are used so that the advantage of the Crank–Nicholson scheme allowing relatively large time steps, is lost. To solve this problem we have to adopt the *predictor-corrector* procedure, which consists in taking the time point in the centre of each time interval and calculating β using H at this point. It requires, of course, double calculation and this is time-consumming. Let us say that the base time node is k. Then at this time the value of β is

$$\beta_{i,k} = \frac{a}{(a + bH_{i,k})^2}$$

Based on this value the time node $k + 1/2$ is analysed so that finally for calculating at $k + 1$ time node from k node one takes the value of β

$$\beta_{i,k+1/2} = \frac{a}{(a + bH_{i,k+1/2})^2}$$

A similar procedure can be adopted for two-dimensional problems solved by the ADI method. Also the problems solved by the finite element method can be treated similarly. The *predictor-corrector* scheme is mostly suggested for solving time-dependent problems by means of the finite element method in the sense of weak formulation or Galerkin's method. The procedure is the same since we deal then with the system of differential equations in time which are being solved with finite difference schemes.

It should be added that the Fröhlich curve shown here is an example only and the same procedure can be used with any other approximation curve. The other way of dealing numerically with non-linearity is to use a system of linear equations, the well-known iterative method. This is mostly used for static problems. Approaching the proper value of μ is then tackled with the normal iterative process. Of course, this can be also used for time-dependence problems but it would require then the iterative process at each time step, which seems to be an enormous task for the computer.

Recently, a new variant of the FEM which allows the analysis of non-linear eddy currents has been worked out (Yamada *et al.* 1991). The method concerns time-periodic process when the spectrum of harmonics for both calculated and exciting magnitudes as well as for B–H curve can be determined. The idea of the method is that the balance of each particular harmonic is analysed, so that the method is called harmonic balance finite element method (HBFEM). Finally, the FEM matrix contains the coefficients resulting from harmonic balance analysis and the results are obtained in a direct way (without iterations). The method is limited to time-harmonic problems only but such problems appear in electrical engineering very often. Thus, the method seems to be of great usefulness in technical applications.

6.2.2 *Boundary element method*

A serious problem appears when dealing with non-linear media by BEM. In the past the opinion existed that these problems are intractable by this method. Recently the opinion changed as to the static fields (without regarding eddy currents). If the direct relation between the equation coefficients and the quantity searched for exists, then Kirchoff's transformation can be used, which finally gives the linear differential equation, and the non-linearity occurs in boundary integral equation (Brebbia *et al.* 1984).

The other method consists in moving the non-linear part from the left-hand to the right-hand side and then the boundary matrix is linear while the right-hand side is iterated. However, eddy current problems are not static problems by any means and what is more the dependence mentioned is not direct as the magnetic vector potential is adopted. However, an attempt has been made to solve the non-linear problem by the BEM (Kost 1989) but the domain integration of the calculated magnitude (or its derivative) during the iterative process is still required. Needless to add that this destroys the great advantage of the BEM, namely the possibility of calculating searched-for magnitudes at a few selected points in a given domain.

In the case of eddy currents the only effective solution is to combine two methods, that is, finite element method in the region of non-linearity and Boundary Element Method elsewhere. But this is not a BEM approach to eddy current problems since the method is then used in a non-conducting area. This is the situation for the time being. As the BEM is still being developed one expects that the problem will find a satisfying solution in the near future. But even if this happens there is no doubt that BEM is not suited to solving problems in materials having non-linearity or non-homogeneity (see Chapter 5).

7

EDDY CURRENTS DUE TO MOTION

Relative motion of magnetic or non-magnetic structures carrying currents gives rise to the production of eddy currents which are either desirable or create losses, and are hence unwanted. In most real devices of this type two main parts can be distinguished:

- a stationary part, usually providing the excitation by electromagnets such as the stator of rotating or linear electrical machines, launchers, pumps, etc. In the magnetic circuits of these parts eddy currents are undesirable as they diminish the effectiveness of the device.
- a moving part where eddy currents are desirable since they create forces according to eqn (2.14). In this part eddy currents are closely related to the operation of the device and play the main role in its performance. Typical examples are the wound or cage rotors of induction motors, the dampers of synchronous machines under asynchronous operation, the projectiles of electromagnetic launchers and railguns, the liquid metals in electromagnetic pumps etc.

The accurate prediction of eddy currents is necessary for the optimization of the performance and cost of these devices. The numerical modelling of such problems can be obtained by different techniques depending on the following criteria:

(1) If the relative motion does not imply any variation in the geometry of the problem, that is, at least one of the parts in relative motion is smooth, then all the quantities can be expressed in only one reference frame and the motion can be considered by means of the velocity term (see eqn (2.11)):

$$\mathcal{V} = \mu\gamma \operatorname{curl}(\boldsymbol{v} \times \boldsymbol{H}) \tag{7.1}$$

It is obvious that using the magnetic vector potential the additional term in the field equation (3.18) is

$$\mathcal{V} = \gamma\boldsymbol{v} \times \operatorname{curl} \boldsymbol{A} \tag{7.2}$$

Making use, in turn, of the stream function u one can write the velocity term to eqn (3.39) as

$$\mathcal{V} = \mu\gamma(H_0 \operatorname{div} \boldsymbol{v} + \boldsymbol{v} \cdot \operatorname{grad}_t H_0) \tag{7.3}$$

The above expression is derived under the assumption that outside the plate the exciting field exsists only.

(2) If the relative motion causes some variation of the geometry, that is, both parts in relative motion are slotted, then two local reference frames fastened to each part are necessary. In such a problem eqn (2.11) is no longer valid. The field quantities are then calculated independently by means of the parabolic or Helmholtz equation.

As to item 1 one can use all the methods described in Chapter 4 adding the additional term representing motion only. However, solving eqn (2.11) is difficult due to this term and then the situation described in item 2 may find an application, even if the moving part is smooth (Burais *et al.* 1984).

As to item 2 a difficulty arises in ensuring the continuity of the physical quantities at the joint of the reference frames during the motion. Several approaches have been developed to fulfil this requirement:

- mixed methods: the space between the parts in relative motion is modelled analytically, by boundary elements, by gap element, by integral equation method, etc.
- mesh distortion: a mesh in the space between the parts in relative motion is distorted; after reaching a critical distortion the mesh in the space between both parts must be redefined.

The problems encountered are specific for the 'eddy current-based' moving devices. All the other problems are typical for all electrical devices which use the eddy current phenomenon, so that they are not repeated here.

Coming back to item 1, that is, to the problems in which the velocity term cannot be omitted, it should be underlined that the presence of this term in a field equation causes some additional difficulties when the numerical model of the domain type is concerned. The inclusion of the velocity term causes non-physical oscillations in the solution, similar to those generated by the time-varying term (see Fig. 4.11). To obtain a non-oscillatory solution one should refine the mesh, which for some problems may lead to large numbers of elements. Song and Ida (1992) suggest using upwinding of the finite elements, which consists in fitting the basis functions to the velocity terms.

In the analysis of moving media there exist more difficult problems, namely those concerning the coupling field and mechanical equations. These problems lie beyond the scope of this book; nevertheless, one should mention them since they are strongly connected with the eddy current analysis. Readers who are interested in the entire analysis of moving media, are advised to refer to the paper by Salon *et al.* (1990), where such an analysis is exhaustively presented.

8

PRE- AND POSTPROCESSING

As the creation of a continuous model without numerical results is only of theoretical value, the numerical methods are of little usefulness if not accompanied by systems which enable data to be introduced so as to obtain the technically important information after the numerical procedure is finished. These systems are commonly recognised as pre- and postprocessing systems.

The question arises if the user of computer software is interested in pre- and processing procedures. Since the real engineering problems are so specific, it would be difficult to have a generalized 'input–output' analysis. Thus, one can say that commercial packages, although prepared in a 'user-friendly' manner, should provide the possibility of the user's intervention at the initial and final step. This is the case in the analysis of eddy currents. Indeed, creating a mesh as well as getting appropriate results require some experience of the problem encountered.

In the following sections the problems connected with pre- and postprocessing systems are pointed out only since they really constitute a subject by themselves and lie beyond the scope of this book.

8.1 Preprocessing

All numerical methods, referred to in Chapter 4, require some kind of discretization. From this point of view we may divide the numerical methods into two groups: domain methods and boundary methods. The first group includes the finite element method, the finite difference method, and the finite sums method while the second group is based on the boundary element method and its mutations. The space which is discretized is defined by the name of the method, that is, in the first case the domain in question while in the second one its boundary. The earliest use of the FDM did not require a very sophisticated division as it was mainly based on orthogonal lines distributed uniformly in the domain in question. Also, the method of finite sums (or collocation method) is based on rectangular cells. Of course, the further development of the methods has introduced more complex division. Nevertheless, the problem of discretization has dramatically gained importance with the advent of the finite element method. This resulted from the fact that the method itself allows us to analyse complex shapes, that is, it acts in a free topology. It is obvious that the hand-made division into elements limited the efficiency of the method. Hence, the researchers' interest was directed towards the problem of mesh generation from the very beginning of the appearance of the method. Nowadays we have very sophisticated systems for setting up FEM meshes which are fully automatic and adaptable to the problem considered. It is obvious that difficulties increase drastically when one deals with three-dimensional cases.

Concerning mesh generation in two-dimensional cases the reader is referred to the book by Lowther and Silvester (1986) where some procedures used in MAGNET package are described. Also, one can find some interesting contributions to the problem in the papers presented at COMPUMAG and INTERMAG conferences (for example, Csendes *et al.* 1985, Penman and Grieve 1987). The three-dimensional meshes are also discussed in papers presented at the above mentioned conferences (for example, Mitchell and Penman 1992, Kost and Jänicke, 1992).

The problem of discretization is essentially easier in the boundary element method. Then one has to discretize the boundary instead of the domain and this decreases the order of the problem. But even for this method a fully-automated system is also needed since the boundary mesh should be adapted to the problem considered (Krawczyk and Wesełucha 1988).

The preprocessing procedures constitute some kind of an art and require a sense of imagination on behalf of their designers.

8.2 Postprocessing

Engineers are not interested in numbers if they do not represent a physical picture. Thus, the numerical results of a certain field calculation should lead to information having a technical significance. In eddy current analysis such an information usually takes the form of:

(1) field plot,
(2) local values of field,
(3) integral values and/or performance parameters.

As to item 1 one should notice that the plot of field (flux lines) is not as informative as it is, for example, in the analysis of the electrical field. Nevertheless, some may want to know the magnetic field distribution and this is relatively easy to obtain in the two-dimensional case in terms of the magnetic or the electric vector potential. Namely, in the two cases the field plot may be obtained by drawing equipotential lines, that is, lines of constant A or u. The latter case gives an image of the current lines. In the case of domain methods these lines are naturally obtained since the values of potentials are given in the whole domain. Thus, for each time step in transient problems the field plot can be drawn. But in quasi-stationary cases the solution is given in complex numbers, so that the question arises as to what the constant potential means. Of course, one can take real and imaginary parts separately and then two field plots will be created. This is, however, not informative enough. A better solution is to transform the complex solution to the time domain and then to draw the field plot for each time instant. In this way an analyst will obtain 'snapshots' of the phenomenon considered within one period. The time solution is as follows:

$$A(t) = \mathrm{Re}(\hat{A}) \sin \omega t + \mathrm{Im}(\hat{A}) \cos \omega t \tag{8.1}$$

where $\hat{A}$ is the phasor of potential A.

Making the field plots for a number of time instants, one can easily evaluate the results.

There is a somewhat worse situation when one deals with the boundary methods. Then the values of potentials are given on the boundary only and to find any equipotential line one needs to calculate a lot of internal values which leads to the loss of an important advantage of the boundary approach. The equipotential line can also be obtained by searching for the points of the same potential on line, starting from the known boundary potential. This method is even more troublesome and time consumming than the previous one but from a computer art point of view it is more elegant.

In three-dimensional cases, either treated by vector $\boldsymbol{A}$ or by vector $\boldsymbol{T}$ it is very difficult to give a useful picture of the field distribution because the field is vectorial. However, there exists the widely used method of arrows and which is used as a pictorial representation. Others prefer to show the field distribution in selected cross-sections using two-dimensional procedures. As an example, the typical three-dimensional field distribution is shown in Fig. 8.1.[1]

In recent years colour displays have been introduced and this created the possibility of showing field or eddy current distribution using a variety of colours (Nakamae *et al.* 1990). The colour graphics systems promise to solve the problem of visualization in three-dimensional eddy current (and other vector fields) analysis in the near future.

Finally, one can conclude the above considerations by saying that visualization of the three-dimensional vector field constitutes a problem in itself and limits the numerical modelling of the three-dimensional eddy current problems.

As to item 2 the local values of field are much more informative since they may help an analyst to predict some hazard points caused by eddy currents and they may also be helpful in evaluating the integral parameters. In the case of using direct variables, that is, vectors $\boldsymbol{H}$ or $\boldsymbol{B}$, the local values are determined automatically while using potentials requires an additional effort of differentiation nature. This is relatively simple in

[1]The figure is taken from the Polish textbook by M. Gramz and M. Ziółkowski (1992). *The calculation of three-dimensional electromagnetic fields.* Politechnika Szczecińska.

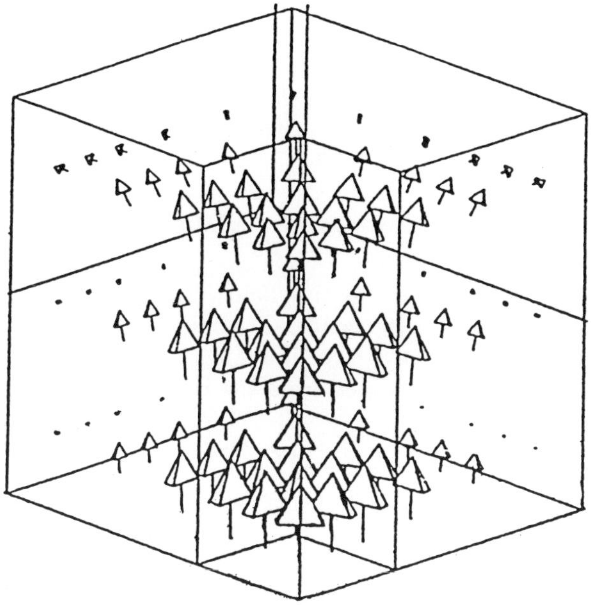

FIG. 8.1. Field distribution

the two-dimensional cases (see eqns (3.7) and (3.8)) and becomes more difficult if the three-dimensional case is considered. But in every case the numerical differentiation is a source of errors and therefore it should be avoided whenever possible.

The problem of calculating local values from potentials is considerably easier when dealing with the boundary element method since the differentiation can be made in a semi-analytical way. This is especially important when one looks for eddy current density in the transient case.

As to item 3 the integral values constitute very often the most important information for an engineer who mainly looks for the two magnitudes:

- forces and related magnitudes (torques, stresses, etc.);
- energy and related parameters (power losses, circuit parameters, etc.).

In Chapter 2, it was shown how to calculate the forces and power losses in the eddy current problems. Let us add here that the Poynting vector may also be used for calculating reactances as far as the quasi-stationary case is considered since magnetic energy is given by the imaginary part of the Poynting vector. As to power losses it should be noticed that sometimes the use of Poynting's vector is not necessary since the local values of eddy current density are given. Then the losses may be calculated as follows:

$$P = \frac{1}{\gamma} \int_V |\, \boldsymbol{j} \,|^2 \, dV \tag{8.2}$$

where V is the eddy current space.

In the quasi-stationary case when stream function $\boldsymbol{u}$ (see section 3.1.2) is applied the eddy current losses can be calculated directly from the function calculated by the following formula:

$$P = -\omega\mu \mathrm{Im} \int_S u H_e^* dS \tag{8.3}$$

where H_e^* is the conjugate to exciting field H_e.

The above formula is very useful as it enables numerical differentiation to be avoided, which is always error-inducing.

The force is usually calculated from the basic formula

$$\boldsymbol{f} = \boldsymbol{j} \times \boldsymbol{B} \tag{8.4}$$

where $\boldsymbol{f}$ is the body force (force density) and the total force results from the integration of $\boldsymbol{f}$.

In some cases it is recommended to use the Maxwell stress tensor to calculate the electromagnetic force, especially when there is no current in the space where the forces exist (for example, air-gap in electromechanical converters). The Maxwell's tensor is directly derived from the above formula and has the following form:

$$\boldsymbol{\tau} = \begin{bmatrix} (B_x^2 - \frac{1}{2} \mid B \mid^2) & B_x B_y & B_x B_z \\ B_y B_x & (B_y^2 - \frac{1}{2} \mid B \mid^2) & B_y B_z \\ B_z B_x & B_z B_y & (B_z^2 - \frac{1}{2} \mid B \mid^2) \end{bmatrix} \tag{8.5}$$

Then the body force is given by

$$\boldsymbol{f} = \frac{1}{\mu} \oint_S \boldsymbol{\tau} \cdot dS \tag{8.6}$$

where S is the boundary of the space considered and dS is taken as the outward normal on it.

The full form is difficult to use but in many technical cases some components of $\boldsymbol{B}$ do not exist and the form becomes simpler. For example, if one component of magnetic flux density exists only as it occurs, for example, in the centre of the air-gap of an electromagnet, the formula derived from the Maxwell's stress tensor is as follows:

$$f = \frac{\mid B \mid^2}{2\mu} \tag{8.7}$$

which is well-known from simpler considerations.

REFERENCES

Antonopoulos, C. S., Tsiboukis, T. D., and Kriezis, E. E. (1992). Field calculation in single- and multilayer coaxial cylindrical shells of infinite length by using a coupled $\boldsymbol{T} - \Omega$ and boundary element method. *IEEE Transactions on Magnetics*, **MAG-28**, (1), 61-8.

Bessho, K., Krawczyk, A., and Nakamoto, K. (1992). New electromagnetic pump based on the effect of magnetic flux concentration. *Compel*, **11**, (1), 149-52

Biro, O. and Preis, K. (1989). On the use of the magnetic vector potential in the finite element analysis of three-dimensional eddy currents. *IEEE Transaction on Magnetics*, **MAG-25**, (4), 3145-59.

Bossavit, A. (1988). A rationale for 'edge elements' in 3-D fields computations. *IEEE Transaction on Magnetics*, **MAG-24**, (1), 74-9.

Bozorth, R. M. (1968). *Ferromagnetism.* van Nostrand, New York.

Brebbia, C. A., Telles, J. C. F., and Wrobel, L. C. (1984). *Boundary element techniques.* Springer, Berlin.

Brebbia, C. A. (ed.) (1990). *Topics in boundary element research.* vol.6: *Electromagnetic applications.* Springer, Berlin

Bunet, P. (1933). *Courants de Foucault.* Baillere et Fils, France.

Burais, N., Foggia, A., Nicolas, A., Pascal, J. P., and Sabonnadiere, J. C. (1984). Numerical solution of eddy currents problems including moving conducting parts. *IEEE Transactions on Magnetics*, **MAG-20**, (5), 1995-7.

Carpenter, C. J. (1977). Comparison of alternative formulations of 3-dimensional magnetic-field and eddy-current problems at power frequencies. *IEE Proceedings*, **124**, (11), 1026-34.

Chari, M. V. K. (1974). Finite element solution of the eddy current problem in magnetic structure. *IEEE Transactions on Power Apparatus and Systems*, **PAS-93**, (1), 62-72.

Chari, M. V. K. and Silvester, P. (eds) (1980). *Finite element method in electrical and magnetic field problems.* Wiley, New York.

Chura, M. (1976). Eddy current transient. *IEE Proceedings*, **123**, (3), 185-9.

Csendes, Z. J., Shenton, D., and Shahnasser, H. (1983). Magnetic field computation using Delunay triangulation and complementary finite element method. *IEEE Transaction on Magnetics*, **MAG-19**, (5), 2551-4.

Davey, K. R. and Barnes, W. J. (1983). Three-dimensional eddy current problems using the $T-\Omega$ method and Fredholm integral equations. *IEEE Transactions on Magnetics*, **MAG-19**, (2), 120-5.

El-Sherbiny, M. K. (1973). Representation of the magnetization characteristic by a sum of potentials. *IEEE Transaction on Magnetics*, **MAG-9**, (1), 60-1.

Emson, C. R. I. and Simkin, J. (1983). An optimal method for 3-D eddy currents. *IEEE Transactions on Magnetics*, **MAG-19**, (2), 2450-2.

Enokizono, T. and Todaka, T. (1990). Three-dimensional boundary element method using a time-dependent solution. *International Journal of Applied Electromagnetics in Materials*, **1**, (2-4), 327-33.

Erdeleyi, E. A. and Fuchs, E. F. (1974). Fields in electrical devices containing soft nonlinear materials. *IEEE Transactions on Magnetics*, **MAG-10**, (4), 1103-8.

Fawzi, T. H., Taher Ahmed, M., and Burke, P. E. (1985). On the use of the impedance boundary conditions in eddy current problems. *IEEE Transactions on Magnetics*, **MAG-21**, (5), 1835-40.

Finlayson, B. (1972). *Variational and weighted residuals methods in chemical engineering*. Academic Press, New York.

Gąsiorski, A. K. (1985). Finite-element method for field analysis in rectangular conductor carrying sinusoidal current. *Archiv für Elektrotechnik*, **68**, (2), 85-92.

Gross H. G. (1940). Die Berechnung der Stromverteilung in zylindriscen Leitern mit rechteckigen und elliptischen Querschnitt. *Archiv für Elektrotechnik*, **34**, (5), 241-57.

Grumet, A. (1959). Penetration of transient electromagnetic fields into a conductor. *Journal of Applied Physics*, **30**, (5), 682-96.

Gurtin, M. E. (1964). Variational principles for linear initial-value problems. *Quart. Appl. Math.*, **22**, 252-6.

Hammond, P. (1971). *Applied electromagnetism*. Pergamon Press, Oxford.

Hammond, P. (1980). *Energy methods in electromagnetism*. Clarendon Press, Oxford.

Hammond, P. (1982). Use of potentials in calculation of electromagnetic fields. *IEE Proceedings*, **129**, Part A, (2), 106-12.

Hammond, P. (1989). Upper and lower bounds in eddy-current calculations. *IEE Proceedings* **136**, Part A, (4), 207-16.

Harrington, R. F. (1968). *Field computation by moment methods*. Macmillan, New York.

Hass, H. (1979). Das Eindrigen eines magnetischen Gleichfeldes in einem ferromagnetischen Halbraum. *Archiv für Elektrotechnik*, **62**, (1), 69-73.

Heller, B. (1953). Prechodne jevy w masiwnym zeleze. *Elektrotechnicky Obzor*, **42**, (7/8), 368-75.

John, F. (1978). *Partial differential equations.* Springer, Berlin.

Kameari, A. (1989). Three dimensional eddy current calculation using edge elements for magnetic vector potential. In: *Applied electromagnetics in materials.* (ed. K.Miya), Pergamon Press, Oxford.

Konrad, A. (1985). Eddy currents and modelling. *IEEE Transactions on Magnetics*, **MAG-21**, (5), 1805-10.

Kost, A. (1989). Calculation of three-dimensional nonlinear eddy currents in an electrical machine by the boundary element method. In: *Boundary element techniques: applications in engineering.* (ed. C.A. Brebbia and N.G. Zamani), Computational Mechanics Publication, Southampton

Kost, A. and Jänicke, L. (1992). Universal generation of an initial mesh for adaptive 3-D finite element method. *IEEE Transactions on Magnetics*, **MAG-28**, (2), 1735-8.

Krakowski, M. (1987). On certain properties of the electric vector potential in eddy-current problems. *IEE Proceedings, Part A*, **134**, (10), 768-72.

Krakowski, M. (1988). Boundary-value problem for eddy currents induced in thin metallic plates by external magnetic field. *Bulletin of the Polish Academy of Sciences, ser. Technical Sciences*, **36**, (7-9), 551-6.

Krawczyk, A. (1980). Unidimensional transient electromagnetic field in a double-layer conductive medium. *Prace Instytutu Elektrotechniki*, **28**, (113), 21-31.

Krawczyk, A. (1983). The application of FEM to transient problems of electromagnetic engineering. *Rozprawy Elektrotechniczne*, **29**, (2), 331-44.

Krawczyk, A. (1986). The calculation of transient eddy currents by means of the boundary element method. In: *Boundary elements VIII* (ed. M.Tanaka and C.A. Brebbia). Computational Mechanics Publication, Springer, Berlin.

Krawczyk, A. (1988). The BEM analysis of transient skin effect in slot of electrical machine. In: *Proceedings on ICEM'88.* Pisa.

Krawczyk, A. (1990). The BEM analysis of transient eddy currents in non-homogeneous regions. In: *Boundary elements in mechanical and electrical engineering.* Computational Mechanics Publications, Southampton.

Krawczyk, A. and Gmitrzak, A. (1992). Electromagnetic forces due to eddy currents. In: *Electromagnetic forces and applications* (ed. J.Tani and T.Takagi). Elsevier, Amsterdam.

Krawczyk, A. and Turowski, J. (1987). Recent development in eddy current analysis. *IEEE Transaction on Magnetics*, **MAG-23**, (5), 3032-7.

Krawczyk, A. and Wesełucha, Z. (1987). BEM software for electromagnetic modelling of electrical devices. *Prace Instytutu Elektrotechniki*, **35**, (147), 111-8.

Krawczyk, A. and Wesełucha, Z. (1992). Indirect calculation of transient eddy currents in special devices. In: *Electromagnetic phenomena and com-*

putational technique (ed. M.Enokizono and J.P.Nowacki). Elsevier, Amsterdam.

Kriezis, E. E. and Xypteras, I. (1979). Eddy current distribution and loss in a semi-infinite conducting space due to a vertical current loop. *ETZ-Archiv*, **100**, (7), 201-7.

Kuhn, T. S. (1962). *The structure of scientific revolutions.* University of Chicago Press, Chicago, Illinois.

Lambert, J. (1973). *Computational methods in ordinary differential equations.* Wiley, New York.

Lameraner, J. and Stafl, M. (1968). *Eddy currents.* Ilife, London.

Lipiński, W. and Krasoń, P. (1981). Stromverteilung in wechselstromdurchflossenen massiven Leitern, berechnet mit Hilfe von Integralgleichung. *Archiv für Elektrotechnik*, **63**, (6), 345-8.

Lowther, D. A., and Silvester, P. (1985). *Computer-aided design in magnetics.* Springer, Berlin.

Mayergoyz, I.D. (1986). 3-D eddy current problems and the boundary integral equation method. In: *Computational electromagnetics*, (ed. Z.J.Csendes), Elsevier, Amsterdam.

Mitchell, A. R. and Wait, R. (1976). *Finite element method in partial differential equations.* Wiley, New York.

Mittra, R. (1973). *Computer techniques for electromagnetics.* Pergamon Press, Oxford.

Mocanu, C. I. (1971). Über die Zeitkonstante des Skineffektes. *ETZ-A*, **92**, (3), 156-61.

Moon, F. C. (1984). *Magneto-solid mechanics.* Wiley, New York.

Müller, W. (1982). A new iteration technique for solving stationary eddy current problem using the method of finite differences. *IEEE Transactions on Magnetics*, **MAG-18**, (2), 588-93.

Nakata, T. and Fujiwara, K. (1992). Present status of 3-D numerical analysis of magnetic fields. In: *Electromagnetic forces and applications* (ed. J.Tani and T.Takagi), Elsevier, Amsterdam.

Nakata, T., Takahashi, N., and Fujiwara, K. (1988). Physical meaning of $\nabla\phi$ in eddy current analysis using magnetic vector potentials. *IEEE Transactions on Magnetics*, **MAG-24**, (1), 178-81.

Nakata, T., Takahashi, N., Fujiwara, K., and Sakaguchi, M. (1990). 3-D open boundary magnetic field analysis using infinite element based on hybrid finite element method. *IEEE Transactions on Magnetics*, **MAG-26**, (2), 368-70.

Pawluk, K. and Szweicer, T. (1978). Biharmonic equation in quasi-stationary field problem. *IEEE Transaction on Magnetics*, **MAG-14**, (5), 563-5.

Penman, J. and Grieve, M. D. (1987). Self-adaptive mesh generation technique for the finite element method. *IEE Proceedings*, **134**, Part A, (8), 561-7.

Potter, D. (1973). *Computational physics.* Wiley, New York.

Rucker, W. M. and Richter, K. R. (1983). Calculation of two-dimensional eddy current problems with the boundary element method. *IEEE Transactions on Magnetics*, **MAG-19**, (6), 2429-32.

Rucker, W. M. and Richter, K. R. (1990). A BEM code for 3-D eddy current calculations. *IEEE Transactions on Magnetics*, **MAG-26**, (2), 462-5.

Salon, J. and Schneider, J. M. (1981). A comparison of boundary integral and finite element formulations of the eddy current problem. *IEEE Transactions on Power Apparatus and Systems*, **PAS-100**, (4), 1473-9.

Salon, S. J., DeBortoli, M. J., and Palma, R. (1990). Coupling of transient fields, circuits, and motion using finite element analysis. *Journal of Electromagnetic Waves and Applications*, **4**, (11), 1077-106.

Schieber, D. (1986). *Electromagnetic induction phenomena.* Springer, Berlin.

Schwenkhagen, H. F. (1927). Untersuchung über Stromverdrängung in rechteckigen Querschnitten. *Archiv für Elektrotechnik*, **17**, (6), 537-589.

Sikora, R., Purczynski, J., Lipinski, W., and Gramz, M. (1978). Use of variational methods to the eddy currents calculation in thin conducting plates. *IEEE Transactions on Magnetics*, **MAG-14**, (5), 383-6.

Silvester, P. and Ferrari, R. L. (1984). *Finite elements for electrical engineers*, Cambridge University Press, Cambridge

Smith, G. D. (1965). *Numerical solution of partial differential equations.* Oxford University Press, London.

Song, H. and Ida, N. (1991). An eddy current constraint formulation for 3D electromagnetic field calculations. *IEEE Transaction on Magnetics*, **MAG-27**, (5), 4012-5.

Song, H. and Ida, N. (1992). Modeling of velocity terms in 3D eddy current problems. *IEEE Transactions on Magnetics*, **MAG-28**, (2), 1178-81.

Stoll, R. L. (1974). *The analysis of eddy currents.* Clarendon Press, Oxford.

Stratton, J. (1941). *Electromagnetic theory.* McGraw-Hill, New York.

Szymański, G. (1982). Electrodynamic forces at thin non-magnetic tapes placed above ferromagnetic plates. *Archiv für Elektrotechnik*, **65**, (3), 179-83.

Tarasiewicz, E., Findlay, R. D., and Dableh, J. H. (1987). The variational treatment of the diffusion equation for vector field problems. *IEEE Transactions on Magnetics*, **MAG-23**, (4), 1972-8.

Tegopoulos, J. A. and Kriezis, M. (1985). *Eddy currents in linear conducting media*, Elsevier, Amsterdam.

Tozoni, O. W. and Mayergoyz, I. D. (1974). *Analysis of 3-D electromagnetic fields.* Technika USSR (in Russian).

Trowbridge C. W. (1985). Low frequency electromagnetic field computation in three dimensions. *Computer Methods in Applied Mechanics and Engineering*, **52**, (3), 653-74.

Tsuboi, H. and Misaki, T. (1987). Three dimensional analysis of eddy current distribution by the boundary element method. *IEEE Transactions on Magnetics*, **MAG-23**, (5), 3044-6.

Wiak, S. (1985). A method of calculation of transient electromagnetic field in nonlinear ferromagnetic medium. *Archiv für Elektrotechnik*, **68**, (6), 113-20.

Wiak, S. and Zakrzewski, K. (1987). Numerical calculation of transients in electrical circuits with nonlinear eddy current skin effect. *IEE Proceedings*, **134**, Part A, (9), 741-6.

Widger, G. F. T. (1969). Representation of magnetization curves over extensive range by rational-fraction approximation. *IEE Proceedings*, **116**, (2), 156-60.

Wilson, E. and Nickel, R. (1966). Application of the finite element method to heat conduction analysis. *Nuclear Engineering and Design*, **4**, 276-286.

Yamada, S., Bessho, K., and Lu, J. (1989). Harmonic balance finite element method to nonlinear AC magnetic analysis. *IEEE Transactions on Magnetics*, **MAG-25**, (4), 2971-3.

Zlamal, M. (1976). Finite element method in heat conduction problems. In: *The Mathematics of finite elements and applications*, Academic Press, New York.

INDEX